AF559755

Marine Geology

Marine Geology

Ajay Mehta

RANDOM PUBLICATIONS
NEW DELHI (INDIA)

Marine Geology

ISBN 978-93-5111-420-8

Published in 2014 in India by

RANDOM PUBLICATIONS

4376-A/4B, Gali Murari Lal, Ansari Road
New Delhi-110 002
Phone : +91-11-43580356, +91-11-23289044
e-mail: randomexports@gmail.com, sales@randompublications.com, info@randompublications.com

Reprinted 2024

Type Setting by: Friends Media, Delhi-110089
Digitally Printed at: Replika Press Pvt. Ltd.

Preface

Marine geology involves geophysical, geochemical, sedimentological and paleontological investigations of the ocean floor and coastal margins. Marine geology has strong ties to physical oceanography. Marine geological studies were of extreme importance in providing the critical evidence for sea floor spreading and plate tectonics in the years following World War II. The deep ocean floor is the last essentially unexplored frontier and detailed mapping in support of both military (submarine) objectives and economic (petroleum and metal mining) objectives drives the research. The Ring of Fire around the Pacific Ocean with its attendant intense volcanism and seismic activity poses a major threat for disastrous earthquakes, tsunamis and volcanic eruptions. Any early warning systems for these disastrous events will require a more detailed understanding of marine geology of coastal and island arc environments. The study of littoral and deep sea sedimentation and the precipitation and dissolution rates of calcium carbonate in various marine environments has important implications for global climate change.

The discovery and continued study of mid-ocean rift zone volcanism and hydrothermal vents, first in the Red Sea and later along the East Pacific Riseand the Mid-Atlantic Ridge systems were and continue to be important areas of marine geological research. The extremophile organisms discovered living within and adjacent to those hydrothermal systems have had a pronounced impact on our understanding of life on Earth and potentially the origin of life within such an environment. Oceanic trenches are hemispheric-scale long but narrow topographic depressions of the sea floor. They also are the deepest parts of the ocean floor. The Mariana Trench (or Marianas Trench) is the deepest known submarine trench, and the deepest location in the Earth's crust itself. A subduction zone where the Pacific Plate is being subducted under the Philippine Sea Plate. The bottom of the trench is further below sea level than Mount Everestis above sea level. Marine geologists approach these problems through studies of oceanic rocks and sediments. Marine geophysicists are primarily concerned with the application of gravity, magnetics, heat flow, and seismic methods to study the structure of the earth beneath the oceans. Within the Joint Program, there are geologists, geochemists, and geophysicists who conduct research on a wide variety of topics

This book offers a comprehensive description of the applications of various fields in this subject. The book will be appropriate as a guide for students.

I thank all members of my team who have helped in the preparation of the book. My special thanks go to "Random Publication" who have published the book.

—Ajay Mehta

Contents

1

Nature of Oceanic Life

Biological Expeditions

Although a number of 16th and 17th century travellers provided much valuable information about the plants and animals in the Orient, America, and Africa, most of this information was collected by curious individuals rather than trained observers. A development that occurred during the 18th and 19th centuries was the organisation of scientific expeditions, usually under the auspices of a particular government. The most notable of these efforts were the voyages of the "Endeavour," the "Investigator," the "Beagle," and the "Challenger," all sponsored by the English government. Captain James Cook sailed the "Endeavour" to the South Sea islands, New Zealand, New Guinea, and Australia in 1768; the voyage provided Joseph Banks, a young naturalist, with the opportunity to make a very extensive collection of plants and notes, which helped establish him as a leading biologist. Another expedition to the same area in the "Investigator" in 1801 included a botanist, Robert Brown, whose work on the plants of Australia and New Zealand became a classic; especially important were his descriptions of how certain plants adapt to different environmental conditions. Brown is also credited with discovering the cell nucleus and analysing sexual processes in higher plants.

One of the most famous biological expeditions of all time was that of the "Beagle" in 1831, the members including Charles Darwin. Although Darwin's primary interest at the time was geology, his visit to the Galápagos Islands aroused his interest in biology and caused him to speculate about their curious insular animal life and the significance of isolation in space and time for the formation of species. During the "Beagle" voyage, Darwin collected specimens of and accumulated copious notes on the plants and animals of South America and Australia, for which he received great acclaim on his return to England. In spite of these expeditions, the contributions made by individuals were still very important. Such an individual was the English naturalist Alfred Russel Wallace, who undertook explorations of the Malay Peninsula from 1854 to 1862. In 1876 he published his book *The Geographical Distribution of Animals*, in which he divided the landmasses into six zoogeographical regions and described their characteristic fauna. Wallace also contributed to the theory of evolution, publishing in 1870 a book expressing his views, *Contributions to the Theory of Natural Selection.*

The Development of the Cell Theory

Although the microscopists of the 17th century had made detailed descriptions of plant and animal structure and though Hooke had coined the term cell for the compartments he had observed in cork tissue, their observations lacked an underlying theoretical unity. It was not until 1838 that Matthias J. Schleiden, a German botanist interested in plant anatomy, stated, "the lower plants all consist of one cell, while the higher ones are composed of (many) individual cells." When Schleiden's friend, the German physiologist Theodour Schwann, extended the cellular theory to include animals, he thereby brought about a rapprochement between botany and zoology. The formation of the cell theory—all plants and animals are made up of cells—marked a great conceptual advance in biology, and it resulted in renewed attention to the living processes that go on in cells. In 1846, after several investigators had described the streaming movement of the cytoplasm in plant cells, Hugo von Mohl, a German botanist, coined the word protoplasm to designate the living substance of the cell. The concept of protoplasm as the physical basis of life led to the development of cell physiology. A further extension of the cell theory was the development of cellular pathology by Rudolf Virchow, who established the relationship between abnormal events in the body and unusual cellular activities. This gave a new direction to the study of pathology and resulted in advances in medicine.

The Theory of Evolution

As knowledge of plant and animal forms accumulated during the 16th, 17th, and 18th centuries, a few biologists began to speculate about the ancestry of these organisms, though the prevailing view was that promulgated by Linnaeus namely, the immutability of the species. Among the early speculations voiced during the 18th century, Erasmus

Darwin, an English physician and the grandfather of Charles Darwin, concluded that species descend from common ancestors and that there is a struggle for existence among animals. A French naturalist, Jean-Baptiste Lamarck, who was probably the most important of the 18th century evolutionists, recognised the role of isolation in species formation; he also saw the unity in nature and conceived the idea of the evolutionary tree.

A complete theory of evolution was not announced, however, until the publication in 1859 of Charles Darwin's In his book Darwin stated that all living creatures multiply so rapidly that, if left unchecked, they would soon overpopulate the world. According to Darwin, the checks on population size are maintained by competition for the means of life. Hence, if any member of a species differs in some way that makes it better fitted to survive, then it will have an advantage that its offspring would be likely to perpetuate. Darwin's work reflects the influence of a British economist, Thomas Robert Malthus, who in 1838 published an essay on population in which he warned that if man multiplies more rapidly than his food supply, competition for existence would result.

Darwin was also influenced by a British geologist, Charles Lyell, who realised from his studies of geological formations that the relative ages of deposits could be estimated by means of the proportion of living and extinct mollusks. But it was not until after his travels in the "Beagle" in 1831, during which he observed a great richness and diversity of island fauna, that Darwin began to develop his theory of evolution. Alfred Russel Wallace had reached conclusions similar to those of Darwin following his studies of plants and animals in the Malay Peninsula.

Conceptually, the theory was of the utmost significance, accounting as it did for the formation of new species. Following the subsequent discovery of the chromosomal basis of inheritance and the laws of heredity, it could be seen that natural selection does not involve the sharp alternatives of life or death but results from the differential survival of variants. Today, the universal principle of natural selection, which is the central concept of Darwin's theory, is firmly established.

The Study of the Reproduction and Development of Organisms

Preformation versus Epigenesis

A question posed by Aristotle was whether the embryo is preformed and therefore only enlarges during development or whether it differentiates from an amorphous beginning. Two conflicting schools of thought had been based on this question: the preformation school maintained that the egg contains a miniature individual that develops into the adult stage in the proper environment; the epigenesis school believed that the egg is initially undifferentiated and that development occurs as a series of steps.

Prominent supporters of the preformation doctrine, which was widely held until the 18th century, included Malpighi, Swammerdam, and Leeuwenhoek. In the 19th century, as criticism of preformation mounted, Karl Ernst von Baer, an Estonian embryologist, provided the final evidence against the theory. His discovery of the mammalian egg and his recognition of the formation of the germ layers out of which the embryonic organs develop laid the foundations of modern embryology.

The Fertilization Process

Despite the many early descriptions of spermatozoa, their essential role in fertilization was not proven until 1879, when Hermann Fol, a Swiss physician and zoologist, observed the penetration of a spermatozoon into an ovum. Prior to this discovery, during the period from 1823 to 1830, the existence of the sexual process in flowering plants had been demonstrated by Giovanni Battista Amici, an Italian astronomer and botanist, and confirmed by others. The discovery of fertilization in plants was of great importance to the development of plant hybrids, which are produced by cross-pollination between different species; it was also of great significance to the studies of genetics and evolution.

The universal occurrence and remarkable similarity of the fertilization process, regardless of the organism in which it occurs, provoked many of the leading investigators of the time to search for the underlying mechanism. It was realised that there must be some way by which the number of chromosomes is reduced before fertilization; otherwise the spermatozoon fused with an egg.

In 1883 Edouard van Beneden, a Belgian cytologist, showed that the eggs and spermatozoa in the worm *Ascaris* contain half the number of chromosomes found in the body cells. To account for the halving of the chromosomes in the sex cells, a process that is called meiosis, in 1887 August Weismann, a German biologist, suggested that there must be two different types of cell division, and by 1900 the details of meiosis had been elucidated.

The Study of Heredity

Pre-Mendelian Theories of Heredity

The fundamental laws of heredity were discovered in 1865 by Gregor Mendel, an Austrian monk and biologist, but his work was ignored until its rediscovery in 1900. There were, however, a number of views on the subject that had been expressed long before Mendel. The Greek philosophers, for example, believed that the traits of individuals were acquired from contact with the environment and that such acquired characteristics could be inherited by offspring.

Because Lamarck was the most famous proponent of the inheritance of acquired characteristics, the theory is called Lamarckism. This concept, which emphasised the use and disuse of organs as the significant factor in determining the characteristics of an individual, postulated that any alterations in the individual could be transmitted to the offspring through the gametes. Yet the inheritance of acquired characteristics has never been experimentally verified, despite many attempts. Furthermore, many of Lamarck's examples, such as the long neck of the giraffe, can be more satisfactorily explained by means of natural selection.

In 1885 Weismann suggested that hereditary characteristics were transmitted by what he called germ plasm—as distinguished from the somatoplasm (body cells)—which linked the generations by a continuous stream of dividing germ cells. In stating definitely seven years later that the material of heredity was in the chromosomes, Weismann anticipated the chromosomal basis of inheritance.

Francis Galton, a 19th century English anthropologist, made a number of important contributions to genetics, one of which was a study of the hereditary nature of ability, from which he developed the concept that judicious breeding could improve the human race (eugenics). Galton's most significant work was the demonstration that each generation of ancestors makes a proportionate contribution to the total makeup of the individual. Thus, he suggested that if a tall man marries a short woman, each should contribute half of the total heritage, and the resultant offspring should be intermediate between the two parents.

Mendelian Laws of Heredity

Tall versus short; round seed versus wrinkled seed. When Mendel fertilised short plants with pollen from tall plants, he found the offspring (first filial generation) to be uniformly tall. But if he allowed the plants of this generation to self-pollinate (fertilize themselves), their offspring (the second filial generation) exhibited the characters of the grandparents in a rather consistent ratio of three tall to one short. Furthermore, if allowed to self-pollinate, the short plants always bred true—*i.e.,* never produced anything but short plants. From these results Mendel developed the concept of dominance, based on the supposition that each plant carried two trait units, one of which dominated the other. Nothing was known at that time about chromosomes or meiosis, yet Mendel deduced from his results that the trait units, later called genes, could be a kind of physical particle that was transmitted from one generation to another through the reproductive mechanism.

Mendel's most important concept was the idea that the paired genes present in the parent separate or segregate during the formation of the gametes. Moreover, in later experiments in which he studied the inheritance of two pairs of traits, Mendel showed that one pair of genes is independent of another. Thus, the principles of segregation and of independent assortment were established.

Mendel's findings were ignored for 35 years, probably for two reasons. Because the distinguished Swiss botanist Karl Wilhelm von Nägeli failed to recognise the significance of the work after Mendel had sent him the results, he did nothing to encourage Mendel. Nägeli's great prestige and the lack of his endorsement indirectly weighed against widespread recognition of Mendel's work. Moreover, when the work was published, little was known about the cell, and the processes of mitosis and meiosis were completely unknown. Mendel's work was finally rediscovered in 1900, when three botanists independently recognised the worth of his studies from their own research and cited his publication in their work.

Elucidation of the Hereditary Mechanism

By 1901 it was understood how the hereditary units postulated by Mendel are distributed; it was also known that the somatic (body) cells have a double, or diploid, complement of chromosomes, while the reproductive cells have a single, or haploid, chromosome number suggested that each chromosome in a pair can exchange the hereditary factors it carries with those of the other chromosome. At first the U.S. geneticist Thomas Hunt Morgan dismissed this concept, but later, when he found that it agreed with his own laboratory findings, Morgan and his collaborators assigned the hereditary units (genes) specific positions, or loci, within the chromosomes. With the genes established as the carriers of hereditary traits, William Bateson, an English biologist, coined the name genetics for the experimental study of heredity and evolution.

Biology in the 20th Century

Just as the 19th century can be considered the age of cellular biology, the 20th century has been characterised by developments in molecular biology.

Important Conceptual Developments

By utilising modern methods of investigation, such as X-ray diffraction and electron microscopy, to explore levels of cellular organisation beyond that visible with a light microscope—*i.e.,* the ultrastructure of the cell—new concepts of cellular function have been produced. Not only has the study of the molecular organisation of the cell probably had the greatest impact upon biology during the 20th century but it also has led directly to the convergence of many different scientific disciplines in order to acquire a better understanding of life processes.

Another 20th century development has been the realisation that man is as dependent upon the Earth's natural resources as are other animals. The progressive destruction of the environment can be attributed, in part, to an increase in population pressure as well as to certain technological advances. Thus, though lifesaving advances in medicine have resulted in a dramatic drop in the death rate, they have also been a factor contributing to the explosive increase in the human population. Moreover, chemical contaminants being introduced into the environment by manufacturing processes, pesticides, automobile emissions, and other means are seriously endangering all forms of life. It is for these reasons that biologists are beginning to pay much greater attention to the relationships of living things to each other as well as to their biotic and abiotic environments.

Intradisciplinary Work

There are many important categories in the biological sciences. Botany, zoology, and microbiology deal with types of organisms and their relationships with each other. Such disciplines are subdivided into more specialised categories; for example, ichthyology is the study of fishes, algology the study of algae. All of them draw upon paleontology, taxonomy, morphology, and evolution. In the past few decades, many developments in physiology and embryology have resulted from studies in cell biology, biophysics, and biochemistry.

This has given rise to cell physiology, cytochemistry, and ultrastructural studies, which aim at correlating structure with function. Ecology, the study of the relations of a group of organisms to its environment, includes both the physical features of the environment and other organisms that may compete for food and shelter. Ecology may be subdivided according to the environment—for example, freshwater ecology and marine ecology—and draws upon animal behaviour. One aspect of cell biology, formerly called cytology, is the investigation of the structure, composition, and function of cells; biochemistry and biophysics provide important information.

Thus, biology encompasses a number of disciplines; in fact, it has become common to divide biology into its several levels of organisation rather than separating the disciplines. It is useful, for example, to differentiate between organismic biology, the study of the whole organism, and cell biology. Similarly the technological advances of the 20th century have allowed increased understanding of the molecules comprising living things and their aggregation and organisation into such structures as chromosomes and membranes. Knowledge of this aspect, called molecular biology, represents the molecular level of organisation. The fourth level, population biology, involves the complex interaction of population of animals and plants with the environment.

Relations with Other Disciplines

In the 17th century, with the invention of the microscope, which made possible study of the cellular level of organisation, biology began to receive the benefits of scientific developments in physics. In the 18th century such developments in chemistry as a better understanding of the nature of oxygen, carbon dioxide, and water began to have important implications for biology. Today, through the disciplines of biochemistry and biophysics, both chemistry and physics have continued to make significant contributions to biology, particularly in the area of molecular biology.

Biology is also very closely related to the disciplines of medicine and agriculture, out of which it developed as an independent discipline. In a sense, the roles have been reversed in the 20th century, for it is basic research being conducted in biology that is contributing to major advances currently being made in medicine and agriculture. It was biological

Another scientific discipline, that of geology, is closely related to the biological study of paleontology. The technique of radiocarbon dating, which was developed by chemists to determine the age of biological remains, has been of great use in the fields of archaeology and anthropology as well as biology. A new discipline, space biology, has arisen through the activities of the scientists and engineers concerned with the exploration of space.

The conceptual framework of biology has had to be altered to accommodate newly discovered facts. In the process biology has received contributions from and made contributions to many other disciplines, in the humanities as well as in the sciences.

Changing Social and Scientific Values

The biologist's role in society as well as his moral and ethical responsibility in the discovery and development of new ideas has led to a reassessment of his social and scientific value systems. A scientist can no longer ignore the consequences of his discoveries; he is as concerned with the possible misuses of his findings as he is with the basic research in which he is involved.

This emerging social and political role of the biologist and all other scientists requires a weighing of values that cannot be done with the accuracy or the objectivity of a laboratory balance. As a member of society, it is necessary for a biologist now to redefine his social obligations and his functions, particularly in the realm of making judgments about such ethical problems as man's control of his environment or his manipulation of genes to direct further evolutionary development.

The Impact of Microbes on the Environment and Human Activities

Beneficial Effects of Microorganisms

Microbes are everywhere in the biosphere, and their presence invariably affects the environment that they are growing in. The effects of microorganisms on their environment can be beneficial or harmful or inapparent with regard to human measure or observation. Since a good part of this text concerns harmful activities of microbes (i.e., agents of disease) this chapter counters with a discussion of the beneficial activities and exploitations of microorganisms as they relate to human culture. The beneficial effects of microbes derive from their metabolic activities in the environment, their associations with plants and animals, and from their use in food production and biotechnological processes.

Nutrient Cycling and the Cycles of Elements that Make Up Living Systems

At an elemental level, the substances that make up living material consist of carbon (C), hydrogen (H), oxygen (O), nitrogen (N), sulfur (S), phosphorus (P), potassium (K), iron (Fe), sodium (Na), calcium (Ca) and magnesium (Mg). The primary constituents of organic material are C, H, O, N, S, and P. An organic compound always contains C and H and is symbolised as CH_2O (the empirical formula for glucose). Carbon dioxide (CO_2) is considered an inorganic form of carbon.

The most significant effect of the microorganisms on earth is their ability to recycle the primary elements that make up all living systems, especially carbon (C), oxygen (O) and nitrogen (N). These elements occur in different molecular forms that must be shared among all types of life. Different forms of carbon and nitrogen are needed as nutrients by different types of organisms. The diversity of metabolism that exists in the microbes ensures that these elements will be available in their proper form for every type of life. The most important aspects of microbial metabolism that are involved in the cycles of nutrients are discussed below.

Primary production involves photosynthetic organisms which take up CO_2 in the atmosphere and convert it to organic (cellular) material. The process is also called CO_2 fixation, and it accounts for a very large portion of organic carbon available for synthesis of cell material. Although terrestrial plants are obviously primary producers, planktonic algae and cyanobacteria account for nearly half of the primary production on the planet. These unicellular organisms which float in the ocean are the "grass of the sea", and they are the source of carbon from which marine life is derived.

NASA receives data from the Terra and Aqua satellites which measures net primary productivity on Earth. These false colour maps represents the rate at which photosynthetic organisms absorb carbon out of the atmosphere. The yellow and red areas show the highest rates, ranging from 2 to 3 kilograms of carbon taken in per square metre per year. The green, blue, and purple shades show progressively lower productivity. Tropical rain forests are generally the most productive places on Earth. However, primary productivity near the seaï¿½s surface over such a widespread area of the Earth makes the ocean roughly as productive as the land.

Decomposition or biodegradation results in the breakdown of complex organic materials to forms of carbon that can be used by other organisms. There is no naturally-occurring organic compound that cannot me degraded by some microbe, although some synthetic compounds such as teflon, styrofoam, plastics, insecticides and pesticides are broken down slowly or not at all. Through the metabolic processes of fermentation and respiration, organic molecules are eventually broken down to CO_2 which is returned to the atmosphere.

Waste management, whether in compost, landfills or sewage treatment facilities, exploits activities of microbes in the carbon cycle. Organic (solid) materials are digested by microbial enzymes into substrates that eventually are converted to a few organic acids and carbon dioxide.

Nitrogen fixation is a process found only in some bacteria which removes N_2 from the atmosphere and converts it to ammonia (NH_3), for use by plants and animals. Nitrogen fixation also results in replenishment of soil nitrogen removed by agricultural processes. Some bacteria fix nitrogen in symbiotic associations in plants. Other Nitrogen-fixing bacteria are free-living in soil and aquatic habitats

Some habitats like this cactus community in the Sonoran Desert, rely on nitrogen-fixing bacteria at the base of the food chain as the source of nitrogen for maintenance of cell material. Every plant in

Oxygenic photosynthesis occurs in plants, algae and cyanobacteria. It is the type of photosynthesis that results in the production of O_2 in the atmosphere. At least 50 percent of the O_2 on earth is produced by photosynthetic microorganisms (algae and cyanobacteria), and for at least a billion years before plants evolved, microbes were the only organisms producing O_2 on earth. O_2 is required by many types of organisms, including animals, in their respiratory processes.

The cyanobacterium, *Synechococcus*, is a primary component of marine and freshwater plankton and microbial mats, The unicellular procaryote is involved in primary production, nitrogen fixation and oxygenic photosynthesis and thereby participates in the cycles of carbon, nitrogen and oxygen. *Synechococcus* is among the most important photosynthetic bacteria in marine environments, estimated to account for about 25 percent of the primary production that occurs in typical marine habitats.

Biogeography, Ecology, and Vulnerability of Chemosynthetic Ecosystems in the Deep Sea

This chapter is based upon research and findings relating to the Census of Marine Life ChEss project, which addresses the biogeography of deep-water chemosynthetically driven ecosystems This project has been motivated largely by scientific questions concerning phylogeographic relationships among different chemosynthetic habitats, evidence of conduits and barriers to gene flow among those habitats, and environmental factors that control diversity and distribution of chemosynthetically driven fauna. Investigations of chemosynthetic environments in the deep sea span just three decades, owing to their relatively recent discovery.

Despite the excitement of many discoveries in the deep ocean since the early nineteenth century, nothing could have prepared the scientific community for the discovery made in the late 1970s, which would challenge some fundamental principles of our understanding of life on Earth. Deep hot water venting was observed for the first time in 1977 on the Galápagos Rift, in the eastern Pacific. To the astonishment of the deep-sea explorers of the time, a prolific community of bizarre animals were seen to be living in close proximity to these vents (Corliss et al. 1979). Giant tubeworms and huge white clams were among the inhabitants, forming oases of life in the otherwise apparently uninhabited deep seafloor (Figs. 9.1A, B, and C). Most of the creatures first observed on vents were totally new to science, and it was a complete mystery as to what these animals were using for an energy source in the absence of sunlight and in the presence of toxic levels of hydrogen sulfide and heavy metals.

Chemosynthetic Ecosystems: Where Energy from the Deep Seabed is the Source of Life

Until the discovery of hydrothermal vents, benthic deep-sea ecosystems were assumed to be entirely heterotrophic, completely dependent on the input of sedimented organic matter produced in the euphotic surface layers from photosynthesis (Gage 2003) and, in the absence of sunlight, completely devoid of any in situ primary productivity. The deep sea is, in general, a food-poor environment with low secondary productivity and biomass. In 1890, Sergei Nikolaevich Vinogradskii proposed a novel life process called chemosynthesis, which showed that some microbes have the ability to live solely on inorganic chemicals. Almost 90 years later the discovery of hydrothermal vents provided stunning new insight into the extent to which microbial primary productivity by chemosynthesis can maintain biomass-rich metazoan communities with complex trophic structure in an otherwise food-poor deep sea (Jannasch & Mottl 1985). Hydrothermal vents are found on mid-ocean ridges and in back arc basins where deep-water volcanic chains form new ocean floor (reviewed by Van Dover 2000; Tunnicliffe et al. 2003).

The super-heated fluid (up to 407 °C) emanating from vents is charged with metals and sulfur. Microbes in these habitats obtain energy from the oxidation of hydrogen, hydrogen sulfide, or methane from the vent fluid. The microbes can be found either suspended in the water column or forming mats on different substrata, populating seafloor sediments and ocean crust, or living in symbiosis with several major animal taxa (Dubilier et al. 2008; Petersen & Dubilier 2009). By microbial mediation, the rich source of chemical energy supplied from the deep ocean interior through vents allows the development of densely populated ecosystems, where abundances and biomass of fauna are much greater than on the surrounding deep-sea floor.

Eight years after the discovery of hydrothermal vent communities, the first cold seep communities were described in the Gulf of Mexico (Paull et al. 1984). Cold seeps occur in both passive and active (subduction) margins. Seep habitats are characterised by upward flux of cold fluids enriched in methane and often also other hydrocarbons, as well as a high concentration of sulfide in the sediments (Sibuet & Olu 1998; Levin 2005). The first observations of seep communities showed a fauna and trophic ecology similar to that of hydrothermal vents at higher taxonomical levels (Figs. 9.1D and E), but with dissimilarities in terms of species and community structure.

The energetic input to chemosynthetic ecosystems in the deep sea can also derive from photosynthesis as in the case of large organic falls to the seafloor, including kelp, wood, large fish, or whales. After a serendipitous discovery of a whale fall in 1989, the first links between vents, seeps, and the reducing ecosystems at large organic falls were made (Smith & Baco 2003). Bones of whales consist of up to 60% lipids that, when degraded by microbes, produce reduced chemical compounds similar to those emanating from vents and seeps(Treude et al. 2009). Another deep-water reducing environment is created where oxygen minimum zones (OMZs, with oxygen concentrations below 0.5 ml l^{-1} or 22 μM) intercept continental margins, occurring mainly beneath regions of intensive upwelling (Helly & Levin 2004) (Figs. 9.1G and H). Only in the second half of the twentieth century was it understood that OMZs support extensive autotrophic bacterial mats (Gallardo 1963, 1977; Sanders 1969; Fossing et al. 1995; Gallardo & Espinoza 2007) and, in some instances, fauna with a trophic ecology similar to that of vents and seeps (reviewed in Levin 2003).

Adaptations to an "Extreme" Environment

Steep gradients of temperature and chemistry combined with a high disturbance regime, caused by waxing and waning of fluid flow and other processes during the life cycle of a hydrothermal vent, result in low diversity communities with only a few mega- and macrofauna species dominating any given habitat (Van Dover & Trask 2001; Turnipseed et al. 2003; Dreyer et al. 2005). The proportion of extremely rare species (fewer than five individuals in pooled samples containing tens of thousands of individuals from the same vent habitat) is typically high, in the order of 50% of the entire species list for a given quantitative sampling effort (C.L. Van Dover, unpublished observation).

Deep-water chemosynthetic habitats have also been shown to have a high degree of species endemicity in each habitat: 70% in vents (Tunnicliffe et al. 1998; Desbruyères et al. 2006a), about 40% in seeps both for mega epifauna (Bergquist et al. 2005; Cordes et al. 2006) and macro infauna (Levin et al. 2009a). In OMZs, the percentage of endemism is relatively low (Levin et al. 2009a), but has yet to be quantified. Some of the most conspicuous of the endemic species of reducing environments have developed unusual physiological adaptations for the extreme environments in which they live.

These include symbiotic relationships with bacteria, organ and body modifications, and reproductive and novel adaptations for tolerating thermal and chemical fluctuations of great magnitude. Because chemosynthetic habitats are naturally fragmented and ephemeral habitats, successful species must also be specially adapted for dispersal to and colonisation of isolated "chemosynthetic islands" in the deep sea (Bergquist et al. 2003; Neubert et al. 2006; Vrijenhoek 2009a).

Chemosynthetic Islands: A Biogeographic Puzzle with Missing Pieces

Since their discovery just over 30 years ago, more than 700 species from vents (Desbruyères et al. 2006a) and 600 species from seeps have now been described and are listed on ChEssBase (Ramirez-Llodra et al. 2004; This rate of discovery is equivalent to one new species described every two weeks, sustained over approximately one-quarter of the past century (Lutz 2000; Van Dover et al. 2002). Furthermore, geomicrobiologists have explored microbial diversity of chemosynthetic ecosystems, revealing a plethora of interesting and novel metabolisms, but also signature compositions for the different types of reduced habitat, and symbiotic organism (Jørgensen & Boetius 2007; Dubilier et al. 2008).

Although several hundred hydrothermal vent and cold seep sites have now been located worldwide, only approximately 100 have been studied so far with respect to their faunal and microbial composition, and even for their ecosystem function. Nevertheless, through such investigations, scientists soon noticed the differences and in some cases similarities among the animal communities from different vent and seep sites.

For example, why is the giant tubeworm *Riftia pachyptila* only found at Pacific vents whereas shrimp species in the genus *Rimicaris* are only found at Atlantic and Indian Ocean vents? Why is the mussel genus *Bathymodiolus* generally widespread at vents and seeps but largely absent from seeps and vents in the northeastern Pacific Ocean? In 2002, at the onset of the ChEss project, biological investigations of known vent sites provided enough data to describe six biogeographic provinces for vent species (Van Dover et al. 2002) and identified several gaps that needed to be closed to complete the "biogeographical puzzle of seafloor life" (Shank 2004). In contrast, cold seep and whale fall communities appear to share many of the key taxa across all oceans. The ChEss project developed a major exploratory program to address and explain global patterns of biogeography in deep-water chemosynthetic ecosystems and the factors shaping them.

Technological Developments for Exploration

One of the most significant advances in deep-sea investigations of chemosynthetic ecosystems, developed and implemented as a new international state of the art technique within the lifetime of the ChEss project, has been the use of deep-sea autonomous underwater vehicles (AUVs) to trace seafloor hydrothermal systems to their source or to map cold seep systems in the necessary resolution to quantify the distribution of chemosynthetic habitats. This approach (Baker et al. 1995; Baker & German 2004; Yoerger et al. 2007) was sufficient for geological investigations of global-scale heat-flux and chemical discharge to the oceans. However, the ChEss hypotheses concerning global-scale biogeography required more precise location of hydrothermal venting and hydrocarbon seepage on the seafloor; ideally with preliminary characterisation of not only the vent and seep site itself but also a first-order characterisation of the dominant species present. So far, the method has been applied on seven separate hydrothermal vent cruises, from 2002 to 2009, throughout the Southern hemisphere, the least explored part of the global deep ocean. These expeditions have located 16 different new sites on the Galápagos Rift (Shank et al. 2003), in the Lau Basin (southwest Pacific; German et al. 2008a), the Mid-Atlantic Ridge (MAR) (South Atlantic; German et al. 2008b; Melchert et al. 2008; Haase et al. 2009), the southwest Indian Ridge (Southern Indian Ocean; C. Tao, personal communication), the East Pacific Rise (southeast Pacific; C. Tao, personal communication), and the Chile margin (C. German, unpublished observation). For cold seep mapping, a major success was the combined AUV and remotely operated vehicle (ROV) deployment in the Nile Deep Sea Fan, leading to the description of several new types of hydrocarbon seep in depths between 1,000 and 3,500 m (Foucher et al. 2009; technical details described in Dupré et al. 2009).

The way the AUV technique works for the exploration of vents is described in detail by German et al. (2008a). Perhaps most surprising to us, and of widest long-term significance, is that, when flying close to the seafloor, the techniques have not only been sufficiently sensitive to locate high-temperature "black-smoker" venting, but also sites of much more subtle lower-temperature diffuse flow (Shank et al. 2003). Building on these successes, future investigations will be reliant upon the new generation of exploratory vehicles such as a new hybrid AUV–ROV vehicle (Bowen et al. 2009), which has already been applied in ChEss studies as a technological precursor to future under-ice investigations (Jakuba et al. 2008; German et al. 2009).

Finding New Species

In the past decade, we have seen a significant increase in molecular tools for studies to understand species evolution, metapopulations, and gene flow in chemosynthetic regions (Shank & Halanych 2007; Johnson et al. 2008; Plouviez et al. 2009; Vrijenhoek 2009b). New high-resolution and high-throughput methods will result in the first insight into the structure and biogeography of microbial communities of chemosynthetic ecosystems in the Census International Census of Marine Microbes (ICoMM) project. However, a major concern today for marine biodiversity analysis is the paucity of taxonomists using morphological methods, and in particular taxonomists specialising in deep-sea species. Both morphological and molecular taxonomy are essential to develop fundamental knowledge and sustainable management of our marine resources. In an effort to raise the profile of taxonomy once more, ChEss set up an annual program of Training Awards for New Investigators (TAWNI).

These awards have been made to a total of 10 scientists from around the globe to develop further their taxonomic skills relating to chemosynthetic organisms). As a result, they have collectively achieved impressive outputs where many meio-, macro-, and megafauna species have been described and new records identified from different sites. These descriptions have been added to the approximately 200 species that have been described and published from vents, seeps, and whale falls since the onset of the ChEss project in 2002. One of the most extraordinary animals that has consequently received much media attention was discovered on southeast Pacific vents in 2005: the yeti crab *Kiwa hirsuta*. This is not only a species new to science, but also represents a new genus and new family (Macpherson et al. 2005). Recently, a close relative of the vent yeti crab was discovered from Costa Rican cold seeps and is being described with the aid of a TAWNI grant (A. Thurber, personal communication).

Global Biogeography Patterns in Deep-Water Chemosynthetic Ecosystems

Addressing global biogeographic patterns for species from all deep-water chemosynthetic ecosystems and the phylogenetic links among habitats needed a coordinated international effort, with shared human and infrastructure resources, that no single nation could attempt alone. In 2002, ChEss outlined a field program for the strategic exploration and investigation of chemosynthetic ecosystems in key areas that would provide essential information to close some of the main gaps in our knowledge (Tyler et al. 2003).

The ChEss field program was motivated by three scientific questions. (1) What are the taxonomic relationships among different chemosynthetic habitats? (2) What are the conduits and barriers to gene flow among those habitats? (3) What are the environmental factors that control diversity and distribution of chemosynthetically driven fauna? To address these questions at the global scale, four key geographic areas were selected for exploration and investigation: the Atlantic Equatorial Belt (AEB), the New Zealand Region (RENEWZ), the Polar Regions (Arctic and Antarctic), and the southeast Pacific off Chile region (INSPIRE). Below, we describe the issues addressed and main findings in each area.

The Atlantic Equatorial Belt: Barriers and Conduits for Gene Flow

The AEB is a large region expanding from Costa Rica to the West Coast of Africa that encloses numerous seep (for example Costa Rica,

Gulf of Mexico, Blake Ridge, Gulf of Guinea) and vent (e.g., northern MAR (NMAR), southern MAR (SMAR), Cayman Rise) habitats. This region is particularly significant for investigating connectivity among populations and species' maintenance across large geographic areas. Potential gene flow across the Atlantic (west to east) is subject to the effects of deep-water currents (Northeast Atlantic deep water), equatorial jets, and topographic barriers such as the MAR. When considering a north–south direction, gene flow along the MAR may be affected by mid-ocean ridge offsets such as the Romanche and Chain fracture zones.

These fracture zones are significant topographic features 60 million years old, 4 km high and 935 km ridge offset, which cross the equatorial MAR prominently, affecting both the linearity of the ridge system and large-scale ocean circulation in this region. North Atlantic Deep Water flows south along the East coasts of North and South America as far as the Equator before being deflected east, crossing the MAR through conduits created by these major fracture zones (Speer et al. 2003). Circulation within these fracture zones is turbulent and may provide an important dispersal pathway for species from west to east across the Atlantic (Van Dover et al. 2002), for example between the Gulf of Mexico and the Gulf of Guinea.

The cold seeps in the Pacific Costa Rican margin were included in this study to address questions of isolation between the Pacific and the Atlantic faunas after the closure of the Isthmus of Panama 5 million years ago. The fauna from methane seeps on the Costa Rica margin, just now being explored, are yielding surprising affinities, which suggests that this site operates as a crossroads. Some animals appear related to the seep faunas in the Gulf of Mexico and off West Africa, whereas others show phylogenetic affinities with nearby vents at 9° N on the East Pacific Rise and with more distant vents at Juan de Fuca Ridge and the Galápagos (L. Levin, unpublished observations). Furthermore, recent investigations have shown (C. German, C.L. Van Dover & J. Copley, unpublished observation) there is active venting in the ultra-slow Cayman spreading ridge in the Caribbean at depths of 5,000 m (CAYTROUGH 1979), and investigations are underway to determine how the animals colonising these vents are related to vent and seep faunas on either side of the Isthmus of Panama. The first plumes were located in November 2009 at depths below 4,500 m, suggestive of active venting, and these plumes were further explored by ChEss scientists in 2010 who located the source of active venting at 5,000 m – the deepest known vent ever found. Exploration on the

MAR has also led to the discovery of the hottest vents (407 °C) (Kochinsky 2006; Kochinsky et al. 2008), as well as another deep vent (4,100 m), named Ashadze (Ondreas et al. 2007; Fouquet et al. 2008).

In the AEB, the connections across the Atlantic have been relatively clearly defined. The seeps of the African margin contain a fauna with very close affinities to the seep communities in the Gulf of Mexico (Cordes et al. 2007; Olu-Le Roy et al. 2007; Warén & Bouchet 2009) and the seep communities on the Blake Ridge and Barbados accretionary wedge.

The communities are dominated by vestimentiferan tubeworms and bathymodioline mussels and the common seep-associated families of galatheid crabs and alvinocarid shrimp. The bathymodioline species complexes on both sides of the Atlantic sort out among the same species groupings within the genus *Bathymodiolus*, with *B. heckerae* from the Gulf of Mexico and Blake Ridge and *Bathymodiolus* sp. 1 from the African Margin in one grouping, and *B. childressi* from the Gulf of Mexico and *Bathymodiolus* sp. 2 from Africa in another group (Cordes et al. 2007). However, our understanding of the biogeographic puzzle beyond this is less clear and requires further investigation (E. Cordes, unpublished observation).

The discovery of vent sites on the southern MAR (Haase et al. 2007, 2009; German et al. 2008b) and the morphological similarity of their fauna to that of NMAR vents, suggest that the Chain and Romanche fracture zones are less of an impediment to larval dispersal than previously hypothesized (Shank 2006; Haase et al. 2007). Further support for unhindered dispersal of vent fauna along the MAR comes from molecular analyses of the two dominant invertebrates of MAR vents, *Rimicaris* shrimp and *Bathymodiolus* mussels. These studies showed recent gene flow across the equatorial zone for these key host species and their symbionts (Petersen & Dubilier 2009; Petersen et al. 2010). In addition to finding known species, new species, including the shrimp *Opaepele susannae* (Komai et al. 2007), have been described, and now more than 17 (morpho-) species have been identified from the SMAR. Many of these have been genetically compared with taxonomically similar fauna on the NMAR and reveal significant genetic divergence among species considered "the same" in both regions (T. Shank, unpublished observation)

New Zealand Region: Phylogenetic Links Among Habitats

The New Zealand region hosts a wide variety of chemosynthetic ecosystems, all in close geographic proximity. During the ChEss/

COMARGE New Zealand field program (RENEWZ), more than 10 new seep sites were discovered off the New Zealand North Island (Baco-Taylor et al. 2009). One of these sites (Builder's Pencil) covers 135,000 m ^{2}, making it one of the largest known seep sites in the world. These initial and ongoing research activities aim at locating the sites, describing their environmental characteristics, investigating their fauna, and determining potential phylogeographic relationships among species from vents, seeps, and whale falls found in close proximity to one another.

In the New Zealand region, sampling and description of chemosynthetic communities is in its infancy. Baco-Taylor et al. (2009) have now provided an initial characterisation of cold seep faunal communities of the New Zealand region.

Preliminary biological results indicate that, although at higher taxonomic levels (family and above) faunal composition of vent and seep assemblages in the New Zealand region is similar to that of other regions, at the species level, several taxa are apparently endemic to the region. Bathymodiolin mussels and an eolepadid barnacle dominate (in number and biomass) at vent sites on the seamounts of the Kermadec volcanic arc. Genetic analysis of mussels from chemosynthetic habitats by Jones et al. (2006) revealed the New Zealand vent mussel *Gigantidas gladius* to be closely related to species from New Zealand and Atlantic cold seeps.

New Zealand vents are often characterised by the barnacle Vulcanolepus osheai (Buckeridge 2000), found at very high densities which is different from those found farther north in the Pacific, and is most similar to an undescribed species found on the Pacific–Antarctic Ridge (Southward & Jones 2003). The most abundant motile species at Kermadec vent sites are caridean shrimp, including two species of endemic alvinocarids (*Alvinocaris niwa, A. alexander*), and one hippolytid (*Lebbeus wera*) (Webber 2004; Ahyong 2009), as well as two species of alvinocarid found elsewhere in the western Pacific (*A. longirostris, Nautilocaris saintlaurentae*; Ahyong 2009).

Only two species of low abundance and sparsely distributed vestimentiferan worms have been sampled so far from Kermadec vent sites (Miura & Kojima 2006). Of these species, *Lamellibrachia juni* has been found elsewhere in the western Pacific, whereas the other species, *Oasisia fujikurai* , is closely related to *O. alvinae* from the eastern Pacific (Kojima et al. 2006). Other species of macro and megafauna found associated with Kermadec vent sites (Glover et al.

2004; Anderson 2006; Schnabel & Bruce 2006; McLay 2007; Munroe & Hashimoto 2008; Buckeridge 2009) suggest that levels of species endemism in the New Zealand region are relatively high, although some species are either closely related to species, or are found, elsewhere in the wider Pacific region. Community-level analysis (an update of the analysis of Desbruyères et al. (2006b)) suggests that although the New Zealand region does apparently contain a vent community with a distinct composition, there is a degree of similarity with communities from elsewhere in the western Pacific (A. Rowden, unpublished observation).

In total, the analysis of samples either compiled or collected as part of the ChEss/COMARGE project provide some support for the hypothesis that the region may represent a new biogeographic province for both seep and vent fauna. However, there is clearly a need for further sampling.

Exploring Remote Polar Regions

The Polar Regions have received an increasing interest in the first decade of the twenty-first century, facilitated by new AUV technologies being developed to work in these remote areas of difficult access caused by ice coverage (Shank 2004; Jakuba et al. 2008). The exploration of the Arctic Ocean revealed, in 2003, evidence for abundant hydrothermal activity on the Gakkel Ridge (Edmonds et al. 2003). The Gakkel Ridge is an ultra-slow spreading ridge, which lies beneath permanent ice cover within the bathymetrically isolated Arctic Basin. The deep Arctic water is isolated from deep-water in the Atlantic by sills between Greenland and Iceland and between Iceland and Norway. This has important implications for the evolution and ecology of the deep-water Arctic vent fauna.

In July/August 2007, the AGAVE (Arctic GAkkel Vent Exploration) project investigated the Gakkel Ridge using AUVs and a video-guided benthic sampling system (Camper). Investigations suggested "recent" and explosive volcanic activity (Sohn et al. 2008). Extensive fields and pockets of yellow microbial mats dominated the landscape. Microbial samples revealed highly diverse chemolithotrophic microbial communities fueled by iron, hydrogen, or methane (E. Helmke, personal communication). Macrofauna associated with these mats included shrimp, gastropods, and amphipods with hexactinellid sponges peripherally attached to "older lavas". These communities may be sustained by weak fluid discharge from cracks in the young volcanic surfaces (T. Shank, unpublished observation).

So far, the northernmost vent sites that have been investigated by ROV are at 71° N on the Mohns Ridge (Schander et al. 2009). The shallow (500–750 m) sites located there support extensive mats of sulfur-oxidising bacteria. However, of the 180 species described from two fields explored, the only taxon that is potentially symbiont-bearing is a small gastropod, *Rissoa* cf. *griegi*, also known from seeps and wood falls in the North Atlantic. Arctic cold seeps have also been investigated at the Haakon Mosby Mud Volcano (HMMV) on the Barents Sea slope (72° N) at 1,280 m depth (Niemann et al. 2006; Vanreusel et al. 2009). This site has large extensions of bacterial mats and is dominated by siboglinid tubeworms (Lösekann et al. 2008), with many small bivalves of the family Thyasiridae living among them. In terms of macrofauna, the HMMV is dominated by polychaetes, with higher abundances and diversity at the siboglinid fields compared to the bacterial mats.

The meiofauna is dominated by benthic copepods in the active centre, whereas the nematode *Halomonhystera disjuncta* dominates in the bacterial mats (Van Gaever et al. 2006). The other Nordic margin cold seeps of the Storegga and Nyegga systems are also characterised by a high abundance of potentially endemic siboglinid tubeworms in association with methane seepage, as well as the occurrence of diverse mats of giant sulfide-oxidising bacteria, attracting large numbers of meio and macrofauna (Vanreusel et al. 2009).

In the Southern Ocean, vent exploration in the East Scotia Arc and seep investigations on the Weddell Sea have addressed the role of the Circumpolar Current in dispersal of deep-water fauna as a conduit between the Pacific and the Atlantic, or as a barrier between these two oceans and the Southern Ocean. The ChEss (ChEss in the Southern Ocean) project explored the East Scotia Ridge in 2009 and 2010, providing further detail to the vent plume data described by German et al. (2000). A follow-up cruise is planned for 2011 to investigate further and locate the vent source and any potential vent fauna.

On the continental margin of the Antarctic Peninsula in the Weddell Sea, cold seep communities were discovered in 800 m water beneath what was the Larsen B ice shelf. The site was once covered by extensive areas of bacterial mat and beds of live vesicomyid clams (Domack et al. 2005), but hydrocarbon seepage appeared extinct only a few years later (Niemann et al. 2009). The discovery of vesicomyid clams is evidence that the hydrographic boundary between the southern Atlantic and Pacific Oceans with Southern Ocean is not a biogeographic barrier, at least for this taxon, though it remains to be determined if the Weddell Sea vesicomyid has been sufficiently isolated to be genetically distinct from any other vesicomyid species. An international team has recently returned to the Larsen B seep sites to determine the phylogeographic alliance of the Weddell Sea clams with other vesicomyids from the Atlantic and Pacific Basins.

Southeast Pacific Off Chile: A Unique Place on Earth

The southeast Pacific region off Chile is of high interest for deep-water chemosynthetic studies, especially for the study of inter-habitat connectivity through migration and colonisation. Only here can we expect to find every known form of deep-sea chemosynthetic ecosystem in very close proximity to one another. A key reason for this unique juxtaposition of chemosynthetic habitats is the underpinning plate-tectonic setting. The Chile Rise is one of only two modern sites where an active ridge crest is being swallowed by a subduction zone and the only site where such subduction is taking place beneath a continental margin (Cande et al. 1987; Bangs & Cande 1997).

Consequently, one would expect to find hydrothermal vent sites along the East Chile Rise, cold seeps associated with subduction along the Peru–Chile trench at the intersection with the Chile Rise, and an oxygen minimum zone that abuts and extends south along the Peru and Chile margins (Helly & Levin 2004). Along with these geologic/oceanographic occurrences, significant whale feeding grounds and migration routes occur on the southwest American margin (Hucke-Gaete et al. 2004), and there is strong potential for wood-fall from the forests of southern Chile as the Andes slope steeply into the ocean south off approximately 45° S (V. Gallardo, personal communication).

To what extent will the same chemosynthetic organisms be able to take advantage of the chemical energy available at all of these diverse sites? Alternatively, will each type of chemosynthetic system host divergent fauna based on additional factors (for example depth, longevity of chemically reducing conditions, extremes of temperature, and/or fluid compositions)? Some seep sites are already known further north along the margin (Sellanes et al. 2004) and first evidence for hydrothermal activity on the medium-fast spreading Chile Rise was suggested by metalliferous input to sediments in this region (Marienfeld & Marching 1992). Systematic exploration at the very intersection of the ridge-crest and adjacent margin has recently been conducted during a joint ChEss–COMARGE cruise (February–March 2010) and sources of venting were recorded, along with evidence of at least one cold seep site relatively close by, thereby confirming expectations of the scientists on board (A. Thurber, personal communication). The Peru–Chile margin and subduction zone contain hydrate deposits and seep sites venting methane-rich fluids (Brown et al. 1996; Grevemeyer et al. 2003; Sellanes et al. 2004). Until recently these habitats had only been sampled remotely by trawl. A recent expedition provided the first Chile seep images and quantitative samples using a video-guided multicorer (A. Thurber, personal communication).

Based on trawl collections, seeps of the Chilean Margin appear to have evolved in relative isolation from other chemosynthetic communities. There are at least eight species of symbiotic bivalves, including vesicomyids, thyasirids, solemyids, and lucinids, and at least one species of the tubeworm *Lamellibrachia* (Sellanes et al. 2008). The bivalve species do not appear to have close affinities to the chemosynthetic fauna of other seeps, in particular the seeps off the coast of Peru (Olu et al. 1996) or New Zealand (Baco-Taylor et al. 2009). The general composition of the community, including a high diversity of vesicomyids (four species) is similar to that of other seep sites of the eastern Pacific. Further taxonomic resolution of this key fauna along with a complete analysis of the Costa Rica fauna will help to refine further the location of this biogeographic puzzle piece.

Larval Ecology: Shaping Faunal Distribution Under Ecological Timescales

The biological communities that inhabit chemosynthetic environments face several challenges that arise from the peculiarities of the habitat. Firstly, relatively few species are specifically adapted to the physical and chemical characteristics of these habitats. Secondly, these habitats are generally ephemeral at decadal scales (with the notable exception of seeps and OMZs) either because they are geologically unstable (vents) or because they are short lived (large organic falls). Thirdly, these habitats are patchy and can be separated by hundreds to thousands of kilometres of habitat unsuitable for the organisms that are adapted to chemosynthetic conditions. An additional challenge is that most of the organisms that inhabit chemosynthetic environments are either sessile (being attached to a substratum) or show limited mobility in their adult life; they rely solely on planktonic propagules, mainly larvae (Mills et al. 2009) to maintain existing populations and to colonise newly opened areas (e.g., after an eruption at a vent or when a whale lands on the ocean floor). Given the patchy distribution and ephemeral nature of their habitat, adaptations during larval life can have pronounced implications for the success of these species. Yet, our knowledge of the larval ecology of these species, and of deep-sea species in general, remains extremely limited (reviewed in Young 2003; Mills et al. 2009).

Most of our current understanding of larval ecology is based on species that inhabit hydrothermal vents, and information from other chemosynthetic habitats is sorely lacking. Larval populations are being increasingly sampled to assess their abundance and distribution relative to the hydrothermal vent where they most likely originated. In general, larvae are found in greater abundance near the ocean floor than near the plume at hundreds of metres above the bottom, suggesting that they may be dispersing along the ocean floor, taking advantage of the along-axis currents there. Although these types of study were initiated in the 1990s (e.g., Kim et al. 1994; Kim & Mullineaux 1998), they accelerated in the 2000s. However, they have only focused at a handful of sites on the Juan de Fuca Ridge (Metaxas 2004), the East Pacific Rise (Mullineaux et al. 2005; Adams & Mullineaux 2008), and the mid-Atlantic Ridge (Khripounoff et al. 2001, 2008).

Similar studies were initiated in the 2000s and are ongoing at vents in Lau Basin and volcanically active seamounts on arcs in the southern (Kermadec) and western (Mariana) Pacific. These studies suggested that hydrodynamics can provide a mechanism of both larval retention to re-seed existing populations, as well as along-axis transport and dispersal to colonise newly opened areas within hundreds of kilometres. Larval abundance in cold seeps has been measured for two species in the Gulf of Mexico (Van Gaest 2006; Arellano 2008), indicating that, unlike most species at vents (except some crustaceans), larval migration from the seep of origin to surface waters likely occurs.

Larval colonisation is better understood and has received more attention than larval dispersal. Since 2002, studies have focused on vents, seeps, whale and wood falls. In all habitats, larvae of different species settle and colonise areas in a particular sequence that appears to be related to chemical and biological cues of the environment.

We now know that colonisation and succession at vents can be quite rapid and communities can recover from catastrophic disturbances within 2–5 years (Shank et al. 1998). It appears that the spatial and temporal patterns of colonists are primarily related to the physicochemical environment and secondarily to biological interactions. However, the evidence on the latter is still scant, and both experimental manipulations and numerical modelling are being used increasingly to address this gap (Neubert et al. 2006; Shea et al. 2008; N. Kelly, personal communication). The largest gap in our understanding of larval life is the factors that affect larval growth and development in all chemosynthetic environments. The main challenge is larval rearing for species that inhabit deep-water environments with very particular chemical and physical characteristics. Only a few studies have succeeded in rearing larvae of only a handful of species from vents (Marsh et al. 2001; Pradillon et al. 2001), seeps (Young et al. 1996; Van Gaest 2006; Arellano & Young 2009), and whale falls (Rouse et al. 2009), but none were successful in following larvae through to the end of that life stage. Larval rearing in situ has been attempted and has been partly successful (Marsh et al. 2001; Pradillon et al. 2001; Brooke & Young 2009).

A key unknown aspect of the question of larval development is the duration of the larval stage, the period that larvae spend in the water column, and thus their potential dispersal distance. Some specific exceptions include experiments to estimate dispersal time in the vent tubeworm *Riftia pachyptila* (Marsh et al. 2001), the whale-fall polychaete *Osedax* (Rouse et al. 2009), the vent gastropod *Bathynerita* (Van Gaest 2006), and the mussel *Bathymodiolus* (Arellano & Young 2009). Other major gaps in knowledge include larval growth rates in relation to temperature, pressure, and food availability; the cues that induce them to stop swimming and settle onto a suitable habitat (or avoid an unsuitable one); and the role of larval behaviour in vertical positioning while in the water column or near bottom. Increased knowledge of larval connectivity and colonisation processes will add conceptual understanding of reducing ecosystems as metapopulations (Leibold et al. 2004; Neubert et al. 2006) and their response or resilience in the face of natural and anthropogenic disturbance (Levin et al. 2009b).

Understanding Remote and Dynamic Ecosystems

Discovery of deep-water chemosynthetically driven communities is relatively recent. Thus the investigation of these habitats has a very strong exploration component. Biologically, the unknowns exceed the knowns. How many species are there? What is their distribution and why? How do species reproduce, disperse, and colonise new sites? What are the evolutionary history and phylogenetic relationships of species from different chemosynthetic habitats? The remoteness and abrupt topography of deep-water chemosynthetic ecosystems make observation, sampling, and experimentation difficult in these habitats.

There is a strong need for international collaboration and sharing of resources, which has been accomplished through projects such as the Census, contributing greatly to our knowledge of chemosynthetically driven ecosystems. The ongoing learning process allows identifying gaps and provides a driver for science to transform the unknowns into knowns (Gomory 1995; Marchetti 1998).

Chemosynthetic research depends on our capacity to sample, observe, and experiment at great depths in "extreme" physicochemical conditions. Therefore research in chemosynthetic ecosystems closely follows technological developments. For example, the use of submersibles and ROVs made possible direct observation and in situ precise sampling of the ecosystem. AUVs are useful for investigation of regions that are not accessible from the surface (i.e. oceans under ice) and in revolutionising the efficiency of deep-sea hydrothermal exploration (German et al. 2008a; Sohn et al. 2008). Deep-towed sidescan sonar instruments can produce detailed acoustic images of the deep seafloor.

Submersible-mounted multibeam bathymetry can achieve centimetre-scale resolution maps of the sea floor. High-resolution, high-definition cameras are being used to produce photo-mosaics of remote habitats, providing a comprehensive overview of the habitat and allowing for first interpretations of the relationships between habitat and fauna. New types of biogeochemical sensor module and incubation instrument allow quantification of the transport of energy and the benthic community activities in situ (Boetius & Wenzhöfer 2009). Technology is also advancing rapidly in laboratory techniques, for example molecular biology, which has aided our understanding of processes that, until recently, were hidden from our senses. However, are there limits to knowledge? This question can be considered in terms of different timescales. At the ecological timescale, one of the major limits to knowledge is imposed by society itself. The first barrier is one of economics and human resources.

In the case of the discrete and dynamic deep-water chemosynthetic ecosystems, some aspects might be unknown and unknowable at any given time. German & Lin (2004) have estimated that, for fast- and intermediate-spreading ridges, there should be a volcanic eruption approximately every 50 years for any given 100 km of ridge section. Slow-spreading ridges may exhibit much greater irregularity (German & Lin 2004). It would take the equivalent of one 1 to 2 year expedition to explore all of the southern MAR from north to south. Taking into account the episodic volcanic activity of mid-ocean ridges, we would expect three or four major new eruptions to have occurred over the time it would take to explore the whole 7,000 km of the MAR. By that reckoning (approximately one new vent area on the MAR each year), our current rate of discovery (nine sites on the northern MAR since 1986) is not even keeping pace with the rate of new production. Here, the limit to knowledge is a consequence of the dynamic characteristics of mid-ocean ridges.

Cold seeps at passive margins are more stable ecosystems than vents (Sibuet & Olu 1998), but those on active margins (e.g., Chile, Costa Rica) are subject periodically to some of the most violently destructive forces on Earth. We can study these systems in their current "dormant" state and investigate their geophysics and geochemistry interactions and associated fauna. However, the episodicity of the great earthquakes (an approximate 100 year cycle predicted for the magnitude 9.5 great earthquake) may render the responses to such events impossible to observe and the extent to which the associated organisms are impacted by or even anticipate major tectonic events unknowable.

Similar issues apply to stochastic, discrete, and ephemeral "habitats" such as whale falls and large organic falls. The life cycle of vents on fast-spreading ridges (i.e. East Pacific Rise) or cold seeps on active margins is equivalent, in time, to the construction of a cathedral in the Middle Ages (approximately 100 years), whereas on slower-spreading ridges they can extend to more than 10,000 years (Cave et al. 2002), longer than the oldest known European prehistoric constructs such as Stonehenge in the southern United Kingdom. These are unknowables for ChEss.

Furthermore, the fauna from deep-water vents, seeps, and whale falls is often new to science. Although the diversity of megafauna is low and the new species are described at a pace that keeps with discovery, this is not the case for the more diverse meiofauna, where up to 90% of the species collected can be new to science. The inability to keep up the rate of taxonomic identification is increasingly affected by the continuous decrease in taxonomic expertise among the new generations of marine biologists.

Marine Biotechnology and Microbiology

We are in charge of graduate school education on microbiology, molecular biology and epizootiology for aquatic organisms. One of our goal is to advance the sciences regarding comprehensive physiology and ecology for aquatic organisms by molecular and cell biological techniques. Another goal is to prevent and control infectious diseases of cultured organisms.

Marine Microbiology

In this field, main subjects are taxonomy, ecology and host-microbes interaction of marine microbes by using physiological, biochemical and molecular biochemical techniques. We are also searching for useful gene resources of marine bacteria as subject of applied Microbiology.

Main researching themes: Identification and classification of microbe of disease associate with marine animals and plants. Searching for symbiotic marine bacteria associated with marine animals. Molecular ecology of pathogenic bacteria and symbiotic bacteria associated with marine animals. Searching for sensing factors possessing pathogenic bacteria and symbiotic bacteria associated with marine animals to host animals.

Searching for valuable enzyme-coding genes from marine environmental DNAs.

Marine Molecular Biology

Laboratory of Maine Molecular Biology deals with genetic engineering for various functional proteins and enzymes from marine animals and plants, and their applications to industrial uses.

Current Research Subjects are as Follows

Isolation and cDNA cloning of industrially useful enzymes, cellulose, alginate lyase, mannanase, protease, from marine invertebrates such as scallop, abalone, and sea urchin.

Mass production of recombinant enzymes of marine organisms by using bacterial and insect cells. Production of protoplasts from brown seaweeds by using polysaccharide-degrading enzymes from marine invertebrates.

Isolation and characterisation of enzymes related to growth and differentiation of brown seaweeds. Study on regulatory mechanism of troponin from scallop adductor muscles.

Aquatic Epizootiology

Outbreaks of diseases are one of the most serious problems on aquacultured organisms for securing of food source. In our laboratory, we focus on infectious diseases of fish and shellfish, especially diagnosis and treatment of the diseases, moreover molecular and microbiological studies on the causative agents of the diseases. Also, we perform studies on prevention of diseases for aquatic organisms.

Marine Bioresource Chemistry

The laboratory consisted of two fields, Marie Biofunctional Chemistry and Marine Bio-analytical Chemistry, is engaged in the education for the graduate school.

The lectures on the understanding of unique life in marine organism, characterisation of unique compounds for marine life, functionalities of newly found compounds from marine organism, and maximal utilisation of marine resources as food, nutraceutical, and pharmaceutical compounds.

Biofunctional Chemistry

Marine resources contain a variety of functional compounds. We study on the chemical properties, nutritional effects, biological functions, health benefits, and utilisation of these marine bioresources, and their application to food materials, nutraceuticals and medicines. Absorption, metabolism, nutrigenomics, and molecular biology of marine functional compounds are also our main research themes.

Our research project also includes stabilisation of functional components in food systems and their enzymatic modification for higher biological activities.

Bio-analytical Chemistry

This field has focused on teaching and researching on analytical chemistry stereochemistry and biochemistry of marine organic compounds, particular lipids and lipid related compounds.

Current research activities: stereochemistry of glycerolipids in fish, algae and bacteria; stereospecificity of lipid related enzymes such as lipases; precision analysis of fish oils; analysis, distribution and function of marine organic compounds with unusual structures.

Biomolacular Chemistry

The field is engaged in the education and research on the understanding of unique properties of various compounds in marine resources, and the development of the purification, analysis separation, isolation, and evaluation systems.

Marine Products and Food Science

This laboratory provides research training for graduate students and also carries out various other activities which include development of new theories and highly improved technologies for effective utilisation of marine bioresources as foods and other value added or highly consumed products.

Graduate students are trained in specialised fields and research along their priorities of interest in interdisciplinary fields including chemistry, biology, biochemistry, microbiology, and also production engineering in the related fields.

Marine Bioresources Utilisation

To explore the useful potentials as health beneficial, drugs and for other purposes materials from marine bioresources are the mission of this field. We are identifying some beneficial enzyme inhibitors such as against glucosidase, urease, glucuronidase and so forth from algae. Our current topic is the discovery of a beneficial therapeutic potential against tumour of marine phospholipid liposomes which act as an immune system booster in combination with glucan.

Food Biochemistry

With a view to utilise the fish and shellfish at a high-degree, we educate and study on the biochemical properties of muscle proteins and endogenous enzymes, such as proteinases and transglutaminases, which are closely related to the freshness and quality of seafood. We also work on the structure-activity relationship of fish and shellfish proteins by using molecular biological and genetic techniques. Based on these studies, we attempt to develop a novel preservation methods and to make new seafood products by control of the endogenous enzymes in fish and shellfish.

Food Functional Chemistry

This field covers the applied science and technology for high utilisation of marine bioresource proteins as food materials and bioproducts. In recent topics are as follows: (1) Applied protein chemistry: Attempts to improve protein-functionality (food process characteristics and biofunctions) using molecular modification. (2) Seafood allergy: Identifying allergenic proteins in seafood and effect of food processing on the allergenicity. (3) Applied marine enzymology: Search for new enzymes and enzyme-inhibitors from unutilised marine bioresources and developing their application.

Food Safety

This field includes food preservation and food hygiene, and deals with following subjects to assure wholesomeness of marine foods: (1) specification of causative microorganisms and chemical substances of food deterioration and food-borne diseases, (2) analyses of mode of the actions and development of controlling methods with , cellular biological and molecular biological techniques, and (3) classification and beneficial utilisation of food microorganisms.

Chair of Marine Biosafety Science and Technology

The total safety of marine organisms from the production process to consumption is an urgent social need. This laboratory is designed to educate students and conduct research on both Marine Environmental Biosafety Management and the Marine Life Safety and Quality Management of Food.

Marine Environmental Biosafety Management

The main objectives of this field are the management of aquaculture environment and production of safe aquaculture feed from fisheries wastes for sustainable aquaculture. We provide education and conduct research on the separation science and technology for the removal of toxic substances from aquatic environments and feed ingredients.

Food Chain Safety and Quality Management

To ensure the safety and reliability on food chain, we have to continue the analysis and control of biological, chemical and physical hazard factors during growth, harvesting, processing, distribution and consumption of food. We try to systematize the safety management of food chain. The other target of this chair is to train persons who have highly expert knowledge about international assurance system of food.

Microbiology

Marine microbiology is the study of marine life forms that cannot be seen with the naked eye, including viruses, bacteria, fungi, protozoa and microscopic algae. Representatives of each group have been found in almost every marine habitat examined, from tropical coral reefs to the Antarctic and the greatest depths of the ocean. Microbes occur in the water column and sediments, on the surfaces of marine plants, animals and inanimate objects, in the intestinal tracts of fishes and invertebrates and even within the tissues of other living things. There is a staggering diversity of forms and modes of life among marine microbes. Some require light for growth, many require the saline nature of sea water, while others can survive without oxygen.

Because of their small size, the importance of microbes to life in the oceans is often not appreciated. Knowledge of their activities assumes a new importance when one considers that each millilitre of sea water may be the home for millions of individuals, and each gram of sediment may support thousands of millions of microbes. Microbes play a crucial role in ocean food webs, which support the diversity of life in the sea. Scientists therefore study the microscopic plants that fix carbon dioxide, the bacteria which grow on the myriad of organic, inorganic, dissolved and particulate chemicals in the sea, and the protozoa which feed on them. In this way knowledge is gained about how each life form contributes to all levels of the food web.

Other microbiologists investigate the unique ways in which microbes change chemicals from one form to another, and the importance of this to the health of the oceans. For example, particular kinds of microbes carry out some steps in the cycling of nitrogen from a dissolved gas in the sea to proteins in plants, animals and microbes, to inorganic forms and finally back to a gas. Maintenance of this cycle is essential for life in the oceans.

Many aspects of marine microbiological research have the potential to provide practical information and to form the basis of new biotechnological developments. Microbiologists studying diseases of marine organisms are providing disease diagnosis and control strategies to those attempting to grow marine animals such as prawns, molluscs and fish. New chemicals with potential industrial applications, such as friction-reducing lubricants on the surfaces of some fish, have been found to be produced by marine microbes. Major efforts are also being put into the search for new pharmaceuticals from marine microbes, such as antibiotics and anticancer agents.

The incredible diversity of microbes, their wide distribution in marine habitats, metabolic capabilities and intimate involvement in processes occurring in the oceans ensures lifetimes of discovery for scientists. Research into the activities of microbes involves opportunities for collaboration with scientists from disparate disciplines such as botany, geology, zoology, chemistry and marine biology. Field work and laboratory analysis using traditional microbiological methods as well as sophisticated techniques such as DNA technology are all tools of trade for the marine microbiologist. Marine microbiologists require tertiary qualifications and are employed by universities, a variety of state and federal departments and instrumentalities, and by private industry.

Improving Human Life, Beginning with the Basics

If we can master the fundamental biological processes that shape life, we will possess the basic information needed to address diseases that afflict individuals at every stage of their lives. The University's initiative, "The Basics of Life," expressly capitalises on this concept. An infusion of private funds will catalyse the investment of others, including government and industry, and enable us to build on our leadership in key areas of basic and clinical research, accelerating the development of lifesaving therapies. In the process, we will establish a powerful model for research universities across the nation.

Creating Knowledge To Save Lives

The premise behind the University of Virginia's initiative, The Basics of Life, is compelling. If we can understand fundamental issues—such as how cells become specialised and form tissue, how they communicate with each other and respond to their environment, and how they migrate through the body—we will gain the insight needed to cure a panoply of diseases that affect human beings.

When lives are in the balance, basic information is crucial. Only through detailed knowledge of the formation, maintenance, degeneration, and regeneration of molecular networks, tissues, and organs can we prevent birth defects; control the rampant tissue growth associated with cancer; slow the deterioration that characterises aging; and facilitate the repair, regeneration, and replacement of injured tissues.

Fundamental advances in biological research promote the development of diagnostic tools and therapies throughout the entire life span. Breakthroughs at any one stage of human development can affect treatment at every other stage. At U.Va., we propose to reorganise our research efforts in new and creative ways so that we can more effectively apply our Basics-of-Life strategy to cure or prevent disease, while developing collaborations and partnerships worldwide to extend this model to other major research universities. We will create University-wide initiatives through three institutes and a research centre: the Morphogenesis and Regenerative Institute, the Virginia Institute for Clinical and Translational Research, the Institute on Aging, and the Pediatrics and Birth Defects Research Centre.

Building from a Position of Strength

Thanks to strategic investments by the University and the Commonwealth during the last five years and the extraordinary success of our faculty in securing outside research funding, we have assembled many of the elements needed to assume national leadership in driving lifesaving breakthroughs in diagnosis and treatment of cancer, birth defects, diabetes, and heart disease through advanced research in biology and medicine.

Morphogenesis and Regenerative Medicine Institute

Founded in 2002, the multidisciplinary Morphogenesis and Regenerative Medicine Institute was formed explicitly to pursue one of the most promising lines of inquiry in biology and medicine: expanding our knowledge of how cells differentiate and organise themselves so that we can harness these insights to regenerate defective or damaged organs. The potential of discoveries in this area is stunning, offering us the ability to repair spinal cord injuries, for instance, or restore lung tissue damaged by years of smoking or coal mining.

The institute brings together U.Va.'s world-leading programs in such fundamental biological processes as the migration of cells through the body, the mechanisms that enable cells to adhere to each other, and the signalling pathways that control the organisation of tissues. It also draws on our outstanding expertise in kidney and blood vessel development, among other areas. Virginia Institute for Clinical and Translational Research.

The newly created Virginia Institute for Clinical and Translational Research will play a critical role in extending the benefits of basics-of-life science to the diseases of adults. This research "translates" basic scientific discoveries into what Mr. Jefferson called "useful science" that has direct applications in patient care. Initially, we intend to combine our expertise in tissue development and diabetes to transform precursor cells in the pancreas to cells that can express insulin. The result of this dramatic breakthrough: a cure for Type 1 diabetes. In the process, we will extend our expertise to other areas, such as vascular disease and cancer, which can benefit from similar approaches.

Institute on Aging

The University's Institute on Aging brings together leading authorities in psychology, biology, medicine, and engineering from the United States and Europe to focus on mobility and independence, cognitive aging, age-related neurological disease, and other important issues. U.Va. scientists are responsible for one of the world's foremost longitudinal studies of aging, allowing medical researchers to compare declines in cognitive function with changes at the cellular and molecular level. These correlations provide the fundamental insights needed to pursue more effective strategies to bolster cognitive function and to prevent and treat such a Ge-related illnesses as Parkinson's and Alzheimer's.

Taking the Next Step

The University has put in place many of the elements needed to realise the promise of its basics of life strategy, and the University's Board of Visitors has committed to strengthen scientific and medical research even further. In recent months, the University has recruited several distinguished senior researchers in neuroscience, cell biology, and epidemiology, all areas that are vital to our knowledge of cell, tissue, and organ development. To complement the exciting work of these institutes, we will establish a Pediatrics and Birthdefects Research Centre that combines

University expertise in the Schools of Medicine, Education, and Law. The Pediatrics and Birth Defects Research Centre will apply insights from basic research to address major threats to the health of the fetus, optimise the care of the newborn, and address the treatment of children with acute brain injury or spinal cord injury. The work of the centre will gain additional momentum from the construction of a new U.Va. Children's Hospital.

There is, nonetheless, much more work to be done. An infusion of significant private funding will jump-start our efforts ($300 million in endowment will be required to support properly the four world-class centres). This support will enable us to make strategic additions to our faculty, develop advanced research instruments and facilities, and assemble multidisciplinary research groups that can speed the translation of basic research into real-world clinical applications. With these new resources, the University of Virginia can become the place where birth defects are prevented, where we learn how to prevent cancer cells from moving to lymphatic nodes, where both Type 1 and Type 2 diabetes are cured, where early heart attacks are prevented, and where cognitive and physical functionality are sustained in old age.

In short, by enabling us to complete and energize this unique network devoted to biology and medicine over the course of the human life, these additional funds will accelerate the flow of lifesaving therapies to those who need them most. Our goal? To improve human life, beginning with the basics.

Molecular Biology

Genome sequence comparison has been an important method for understanding gene function and genome evolution since the early days of gene sequencing. Alignment of DNA sequences is the core process in comparative genomics. In recent years, an important new sequence-analysis task has emerged: comparing an entire genome with another. Several powerful alignment algorithms have been developed to align two or more sequences.

MUMmer

MUMmer is a system for rapidly aligning entire genomes, whether in complete or draft form. MUMmer can also align incomplete genomes; it can handle thousands of contigs from a shotgun sequencing project, and will align them to another set of contigs or a genome using the NUCmer program included within the system. If the species are too divergent for a DNA sequence alignment to detect similarity, then the PROmer program within the environment can generate alignments based upon the six-frame translations of both input sequences. The original MUMmer system, version 1.0, was described in a 1999 Nucleic Acids Research paper. Version 2.1 appeared a few years later and was described in a 2002 Nucleic Acids Research paper, and the most recent version MUMmer 3.0 was described in a 2004 Genome Biology paper.

BLAT

BLAT (The BLAST-Like Alignment Tool) is a new tool for sequence alignment, which is similar in many ways to BLAST. The program rapidly scans for relatively short matches (hits), and extends these into high-scoring pairs (HSPs). However, BLAT differs from BLAST in several significant ways. Specifically, where BLAST builds an index of the query sequence and then scans linearly through the database, BLAT builds an index of the database and then scans linearly through the query sequence. Where BLAST triggers an extension when one or two hits occur in proximity to each other, BLAT can trigger extensions on any number of perfect or near-perfect hits. Where BLAST returns each area of homology between two sequences as separate alignments, BLAT stitches them together into a larger alignment. Both the client/server and the stand-alone can do comparisons at the nucleotide, protein, or translated nucleotide level.

MEGABlast

Mega BLAST uses the greedy algorithm of Zhang et al. for nucleotide sequence alignment search and concatenates many queries to save time spent scanning the database. This program is optimised for aligning sequences that differ slightly as a result of sequencing or other similar "errors". It is up to 10 times faster than more common sequence similarity search and alignment programs and therefore can be used to swiftly compare Microbial population genetics is a rapidly advancing field of investigation with relevance to many areas of science. The subject encompasses theoretical issues such as the origins and evolution of species, sex and recombination. Population genetics lays the foundations for tracking the origin and evolution of antibiotic resistance and deadly infectious pathogens and is also an essential tool in the utilisation of beneficial microbes.

Written by leading researchers in the field, this invaluable book details the major current advances in microbial population genetics and genomics. Distinguished international scientists introduce fundamental concepts, describe genetic tools and comprehensively review recent data from SNP surveys, whole-genome DNA sequences and microarray hybridisations. Chapters cover broad groups of microorganisms including viruses, bacteria, archaea, fungi, protozoa and algae. A major focus of the book is the application of molecular tools in the study of genetic variation. Topics covered include microbial systematics, comparative microbial genomics, horizontal gene transfer, pathogenic bacteria, nitrogen-fixing bacteria, cyanobacteria, microalgae, fungi, malaria parasites, viral pathogens and metagenomics.

An essential volume for everyone interested in population genetics and highly recommended reading for all microbiologists.

The higher taxonomic groups within prokaryotes are presently distinguished mainly on the basis of their branching in phylogenetic trees. In most cases, no molecular, biochemical or physiological characteristics are known that are uniquely shared by species from these groups. Analyses of genome sequences are leading to discovery of novel molecular characteristics that are specific for different groups of bacteria and archaea and provide more precise means for identifying and circumscribing these groups of microbes in clear molecular terms and for understanding their evolution. These new approaches and their limited applications for clarifying microbial systematics are described here. Because of their taxa specificities, further studies on these newly discovered molecular characteristics should lead to discovery of novel biochemical and physiological characteristics that are unique to different groups of microbes.

Microbes are ubiquitous in the world in which we live. With the development of high throughput DNA sequencing technology, there has been an explosion of DNA sequence data on microbes. The major aim of future microbial genomics will be to identify the functional significances of individual gene and genomic fragments and to use the information to help improve human health and promote our society development. One current major undertaking to understand genomic information is the comparative analyses between genomes that are not only distantly related, but also closely related ones.

Such comparative analyses between genomes that have diverged at different evolutionary time scales allow us to extract different types of information about biological functions and evolutionary processes. We review the tools and databases that have been established for comparative analyses of microbial genomes and discuss the implications of such analyses on our understandings of the common properties of life, the extent of genome plasticity and diversity within and between species, the processes and mechanisms underlie the observed genome diversity, and the origin and evolution of life.

Horizontal gene transfer, as a major force in shaping bacterial gene content, has gained incredible attention over the last decade. Along with the fast growing bacterial genome sequence data, there have been an increasingly large number of studies focused on horizontal gene transfer. The studies have been gradually transformed from identifying individual genes that have been horizontally transferred to assessing the general patterns of horizontal gene transfer and evaluating the systematic consequences of massive gene transfers.

The rates of gene transfers have been measured by various methods such as parsimony and maximum likelihood methods. Different phylogenetic methods were applied to a variety of data sets to assess whether there exists a congruent and meaningful bacterial tree. Even though some consensus has been reached, many contradictions have emerged and need to be solved in future studies. Our goal here is to review recent studies on this subject with an emphasis on the emerging patterns in horizontal gene transfer research.

Population genetics examine variation in genes among a group of strains of a particular species. Its major theme is to look at how different environmental factors and selective pressures can affect the distribution of genes and alleles. In this chapter *Yersinia pestis* was employed as an example to illustrate how the techniques are used for population genetic studies and how the achievements of these kinds of studies can be used for rapid identification and tracing the origin of pathogenic bacteria. New emerging techniques, including high throughput sequencing technologies, will give us unprecedented opportunities to understand microevolution and pathogenesis of bacterial pathogens.

Symbiotic nitrogen-fixing Rhizobia are of global significance, both in terms of their ecological relationships and their importance as an environmentally benign source of nitrogen for crop plants. These bacteria are capable of forming mutualistic relationships with a variety of legume hosts, where they convert atmospheric nitrogen (N_2) to ammonia that is used to help meet the nitrogen needs of the host plant. In this chapter, we review our current understanding of the population genetics of this diverse group of bacteria. First we briefly describe the various types of genomic architectures that are found within rhizobial species. Next we outline the phylogenetic relationships among the recognised taxa. We then summarise the results of studies that have examined patterns of molecular genetic variation in rhizobial populations at local, regional, and global levels.

The results of these studies have revealed several important insights, including the existence of extensive diversity within, as well as significant genetic differentiation between local and regional populations. The results also provide evidence for long-distance gene flow between continental populations. Several studies have also indicated the existence of low-to-intermediate levels of recombination within rhizobial populations. Fine-scale studies of specific genomic components (e.g., symbiotic plasmids) have also shown that certain genomic elements appear to be more prone to recombination than others.

The results also showed that the different loci responsible for the development of the symbiosis appear to be under different forms of selection. Because most the population studies to date have focused on strains from root nodules, surprisingly little is known on the population genetics of the more numerous non-nodulating soil rhizobia. Future efforts to characterise these populations should significantly enhance our ability to manipulate rhizobial populations in agricultural ecosystems. Cyanobacteria are a group of ecologically diverse photosynthetic bacteria. Because niche differentiation is ultimately the product of differences among individuals within populations, understanding the evolutionary origins of this diversity ultimately requires a population genetics perspective. This chapter considers the current state of our understanding of the mechanisms that generate variation in cyanobacteria, the distribution of this diversity and its potential functional importance, with an emphasis on recent work from our laboratory regarding diversity within and among populations of the cosmopolitan, multicellular cyanobacterium, *Mastigocladus laminosus*. It concludes with an appraisal of the potential to take a population genomics approach to address fundamental questions regarding the nature of adaptive variation and niche differentiation in these microorganisms.

Algae are a highly diverse group of protists, ranging from simple, unicellular organisms to complex, multicellular entities with a range of differentiated tissues and distinct organs. They are found among diverse aquatic ecosystems and play important roles by supplying carbon and energy as well as providing habitat to other members of the biological communities. Some algae cause significant environmental and health problems. However, despite their importance, relatively little is known about this group of organisms.

The first section of this chapter briefly summarises our current understanding of the diversity and evolution of the algae. The second section reviews the molecular markers used to address algal population genetic issues. The third section summarises the population genetic analyses of three algal groups: the dinoflagellates, the diatoms and the haptophytes. With the application of genomics information and technology, the field of algal population genetics is approaching an exciting period of development and expansion.

Mutualisms are reciprocal exploitations that nonetheless increase the fitness of each interacting partner. Two groups of fungal mutualists of plants, epichloë endophytes of grasses and arbuscular mycorrhizal fungi, were selected as focal systems to discuss population-level processes that contribute to the establishment and maintenance of mutualistic interactions. These two classes of fungal cooperators of plants are subject to different and often conflicting selective pressures and represent distinct trajectories of mutualism evolution. Yet, in both cases population structure of symbionts is a source of information critical for understanding how these fungi interact with their hosts.

This article reviews the more common DNA-based markers that are used to genotype fungi, as well as other eukaryotes, and presents several examples of their application to elucidate the population genetics of mammalian pathogenic fungi. The most common fungal infections are briefly summarised to illustrate the issues that are routinely addressed by population genetics approaches. This exciting area of research is continually growing as the methods become more accessible and medical mycologists recognise the value of studying natural isolates of pathogenic fungi.

Genetic diversity, population structure, and evolutionary history of malaria parasites are among the key factors that will influence our ability to identify genes contributing to drug resistance, parasite development, and disease pathogenesis. These factors also have an impact on vaccine and drug development, parasite source tracking, as well as the formulation of other disease prevention and control measures. For example, a highly polymorphic parasite population will contain ample genetic diversity capable of generating drug resistance genotypes at an accelerated rate; while the presence of homogeneous parasite populations should aid in the development of an effective malaria vaccine. Malaria research in the post-genomic era offers many new tools for use in population genetics analyses.

Many viral pathogens, especially those with an RNA genome, are characterised by their high mutation rates and large population sizes. These features are responsible for the high levels of genetic variation usually found in viral populations and for their rapid response to different selective challenges encountered during their infection and transmission processes. They are quantitatively and qualitatively so different from most other organisms that special models and concepts, such as the quasispecies model, have been developed to better describe the evolutionary dynamics of viral populations. Here, we review these and other salient features of viral pathogens, with a species emphasis on RNA viruses. We describe how population genetics theory provides an adequate framework for analysing and interpreting genetic variation in viral populations, for understanding their dynamics, and to study adaptive processes occurring therein. However, not all the evolutionary changes observed are due to the action of positive selection and this is not an all-mighty agent of evolutionary change. We finish this review by introducing a recently developed framework for integrating the evolutionary and epidemic behaviour of infectious organisms, known as phylodynamics, which is especially well-suited for fast evolving organisms such as viruses. Microbial population genetics examines the spatial and temporal patterns of genetic variation across diverse geographic scales and ecological niches. With the arrival of molecular biological techniques, the past 40 years have seen tremendous progresses in microbial population genetics. However, in recent years, the analyses of genetic materials directly from natural environments have revolutionised our approaches and understandings of the diversity, function, and inter-relationships among microorganisms in diverse natural ecological niches. The emergence and development of this expanding new field, that of metragenomics, has been primarily driven by technical and analytical methods developed from high throughput platforms for cloning, microfluidics, DNA sequencing, robotics, high-density microarrays, 2D-gel electrophoresis, and mass spectrometry as well as associated bioinformatics softwares.

In this chapter, I present a general overview of this exciting development, including the general approaches and common tools used for identifying the diversity and function of microorganisms in natural biological communities. Of special notes are the potential impacts of recent developments in single cell isolation, whole-genome amplification, pyrosequencing, and database warehousing on our understanding of microbial population structures in nature. These exciting developments are bringing significant opportunities as well as new challenges to the field of microbial population genetics.

Epigenetics

Epigenetics is the study of changes in gene expression caused by mechanisms other than changes in the DNA sequence. Epigenetics is a rapidly advancing field with an increasing impact on biological and medical research.

The editors of this book have assembled top-quality scientists from diverse fields of epigenetics to produce a major new volume. Comprehensive and cutting-edge, the 26 chapters in this book constitute a key reference manual for everyone involved in epigenetics, DNA methylation, cancer epigenetics and related fields.

Topics include: early life environment, DNA methylation and behaviour, histone acetyltransferase biology, transgenerational epigenetic inheritance, mammalian X inactivation, epigenetic memory in plants, polycomb group regulation, centromeres and telomeres, DNA sequence contribution to nucleosome distribution, macrosatellite epigenetics, histones, cell-fate specification and reprogramming, DNA methylation in cancer, variant histone H2A and cancer development, RNA modification, paramutation in plants, DNMT3L dependent methylation during gametogenesis, non-coding RNA, bisulphite-enabled technologies, rapid analysis of DNA methylation, microarray mapping, DNA methylation profiling, ChIP-sequencing, genome-wide DNA methylation analysis, and epigenetics in maize. In addition there are useful chapters on bioinformatics in epigenomics, resources and tools for epigeneticists, and educational resources for epigenetics.

This up-to-date reference manual is an essential book for those working in the field and for scientists in other disciplines it represents a major information resource on the fascinating and fast-moving field of epigenetics.

The DNA molecule contains within its chemical structure two layers of information. The DNA sequence that bears the ancestral genetic information and the pattern of distribution of covalently bound methyl groups to cytosines in DNA. While the genetic information is similar in all tissues in the individual, the pattern of distribution of methylation across the genome is cell-type specific. DNA methylation is an important regulator of gene function. Recent data that will be discussed here that supports the hypothesis that DNA methylation is a reversible biological signal.

This expands the potential role of DNA methylation beyond embryogenesis to other time-points in life and to post mitotic tissues such as the brain. DNA methylation is proposed to act as a genomic response to both physical and social signals from the environment at different time points in life and to serve as a genomic memory of these exposures at different time scales, stably altering gene expression programming and thus modulating the physical and behavioural phenotypes to respond to these environments. It is hypothesized that DNA methylation provides within the structure of the DNA a dynamic interface between the changing world around us and the relatively fixed and stable genome.

A histone $(H3\text{-}H4)_2$ tetramer flanked by two H2A-H2B heterodimers form the core protein structure, around which DNA is wrapped. DNA and the histone octamer together form the smallest chromatin particle, the nucleosome. How intimately the DNA associates with the core histones and how tightly the nucleosomes are packed with each other is determined by a key post-translational modification of the histone proteins, namely acetylation. Histone acetylation was first discovered in the early 1960s.

After a dearth of progress, due to technical limitations, our knowledge of histone acetylation has exploded in the last fifteen years. Enzymes that catalyse acetylation of histones, the histone acetyltransferases, have been discovered, proteins associated with these have been identified and their preferences for specific histone residues have been determined. Importantly, we are gaining a better understanding of the relevance of histone acetylation in health and disease through the discovery of genetic mutations underlying human diseases in loci encoding histone acetyltransferases (HATs) and through examination of mouse strains deficient in specific histone acetyltransferases. Here we discuss the principles of histone acetyltransferase biology.

Epigenetic states are faithfully inherited through mitotic cell division, but are generally cleared and reset on passage through the mammalian germline. But this clearing of epigenetic marks is not always complete, leading to transgenerational inheritance of epigenotype. Transgenerational epigenetic inheritance has been demonstrated in several organisms, including mammals, and has been most comprehensively studied in mouse strains carrying variants of the *agouti* (A^{vy}) and *axin* ($Axin^{Fu}$) alleles. The most prominent feature of transgenerational epigenetic inheritance is its non-Mendelian nature: not all offspring that inherit the genetic locus also inherit the parental epigenetic state. Transgenerational epigenetic inheritance is emerging as an important facet of mammalian biology. It may underlie the etiology of human diseases that display complex patterns of inheritance, including diabetes, mental illnesses and autoimmune diseases. As variable epigenetic states can be inherited on an invariant genotype, epigenetic variation may provide a substrate for Darwinian selection that is independent of genetic variation.

X chromosome inactivation is the method of dosage compensation that has evolved to equalise expression of X-linked genes between female (XX) and male (XY) mammals. In somatic cells only one X chromosome is active; the second X in female cells is silenced early during embryonic development, a process that involves the co-ordination of multiple levels of epigenetic regulation to ensure stable chromosome-wide silencing. In this chapter we shall focus on the molecular mechanisms involved in X chromosome inactivation, discuss how the epigenetic marks are believed to elicit stable transcriptional silencing, and why X inactivation represents an excellent model system for studying. Polycomb-group (PcG) complexes are essential regulators of plant development. These multiprotein complexes repress gene expression by establishing and maintaining trimethylation of lysine 27 at histone H3, a modification that is associated with repressive chromatin. Recent studies have indicated that plant PcG complexes regulate key genes involved in responses to low temperature.

Vernalisation is a long-term response to low temperatures whereby plants coordinate their seasonal flowering to occur after winter. In contrast, acclimation of plants to low temperatures, a key step in the establishment of frost tolerance, involves rapid activation of cold-acclimation genes. In this chapter, we describe the dynamics of PcG-mediated gene regulation underlying these two important agronomic traits that are triggered by low temperatures.

In eukaryotes, each chromosome has one centromere and its ends are protected by telomeres. The centromere is a specialised chromosomal locus that directs kinetochore assembly and provides the site for microtubule attachment, allowing accurate chromosome segregation during cell division. Despite the critical role centromeres play, centromeric DNA sequences are highly variable and not conserved between species. Increasing evidence, including the discovery of functional neocentromeres, suggests that centromere identity and function is epigenetically defined through the formation of a specialised chromatin structure.

This chapter reviews recent studies addressing the structural and functional characterisation of centromere chromatin, its assembly and propagation during cell division. Telomeres are specialised nucleoprotein complexes that protect the chromosome ends from degradation. In recent years, it has become increasingly clear that heterochromatic marks at telomeres act as negative epigenetic regulators of telomere elongation, repress recombination events at the telomere and are critical for maintaining telomere structural integrity. Recent research reporting telomeres being transcribed by RNA polymerase II to give rise to TERRA RNA, open up an additional level of regulation at the telomere. This chapter will discuss the links between the epigenetic status of telomeres, telomere function and telomere-length regulation, and the implications on cellular reprogramming, aging and cancer.

DNA in eukaryotes is efficiently and compactly organised into chromatin, the fundamental subunit of which is the nucleosome: approximately 150 bp of DNA spooled 1.65 times around a histone octamer. The location and density of nucleosomes play a role in regulating nuclear processes including transcription, replication, recombination, and repair. Mechanisms acting *in trans*, like ATP-dependent remodelers and cellular memory complexes, as well as *in cis* features intrinsic to the DNA sequence itself regulate the location and density of nucleosomes. Here, we review the three cis acting DNA sequence features that affect the distribution of nucleosomes: (1) two frameworks defining the relationship between the histone octamer and the underlying DNA sequence (nucleosome occupancy and nucleosome position, then statistical positioning and a nucleosome positioning code), (2) the organisation of DNA into the nucleosome core particle, and (3) specific DNA sequence features and DNA templates that promote or inhibit the formation of nucleosomes. We close by describing three computational algorithms trained on DNA sequence that have been used to predict nucleosome position and density. In summary, we hope to draw attention to multiple aspects of DNA sequence that specify organisation of sequence into nucleosomes and influence the distribution of nucleosomes in eukaryotic genomes.

The recent completion of several mammalian genome sequences makes obvious that we share a near-identical collection of genes. What defines us as human must therefore be encoded within regions of the genome where we differ, providing an added level of complexity that probably influences the spatial and temporal expression of genes. Most DNA sequence variation occurs within the repetitive DNA, once called 'Junk DNA' that accounts for at least half of the human genome, and evidence is mounting for its important role in genome function. Although some repeat elements are conserved to some extent between mammals, their precise copy number and genomic location typically are not.

In addition, some repeats are not conserved, including the large tandem repeats. This chapter focuses on two large tandem arrays in the human genome that can adopt quite different chromatin configurations as a result of epigenetic changes; one as a direct consequence of X chromosome inactivation and the other in the context of disease susceptibility. Both cases highlight how alternate packaging of these unusual DNA sequences probably results in differing functions. In each instance, common denominators are the acquisition of the epigenetic organiser protein CTCF and a distinct change in transcripts originating from the array.

In eukaryotes, the genetic material in the form of DNA is wrapped around histone proteins to form nucleoprotein filaments called chromatin. Histones help package the DNA to fit it inside the nucleus of each cell, which in turn regulates access to the genetic information contained within the DNA. Hence, all DNA transactions are likely to be affected by histone metabolism. Eukaryotes carry multiple histones genes that can potentially generate enormous quantities of histone proteins. When present in excess, the positively charged histones can potentially "stick" non-specifically to the negatively charged DNA and adversely affect processes that require access to DNA. Not surprisingly, aberrant histone stoichiometry, chromatin assembly or chromatin structure lead to genomic instability, which is characterised by the increased rate of acquisition of alterations in the genome and is associated with deleterious human conditions such as cancer and aging. Hence, histone synthesis is coupled to ongoing DNA replication and is regulated transcriptionally and post-transcriptionally. To further avoid the deleterious consequences of excess histones, a post-translational regulatory mechanism was described recently whereby excess histones are targeted for degradation by the ubiquitin-proteasome system. In this chapter, we discuss the causes and consequences of excess histone accumulation, as well as the strategies that cells have evolved to deal with them.

Cell-fate specification and stem-cell renewal are fundamental processes in the development of multicellular organisms. In both animals and plants, a key role for transcription factors in these processes has been established, but an accumulating body of evidence indicates that epigenetic regulation also plays a critical role. Once regarded as stable marks, all epigenetic modifications, including DNA and histone methylation, are now known to be reversible, and a cohort of enzymes that add or erase epigenetic marks has been identified. Throughout the life of an organism, the epigenome is dynamically modified, leading in turn to transcriptional changes and ultimately cell-fate specification. Here, I review the recent literature that has shaped this view. Plants are unique among complex organisms in having the ability to generate a whole new organism from a single differentiated cell. How plant cells retain this amazing capability while maintaining their cell identity is a mystery. I introduce a model system for study of cell-fate specification, stem-cell renewal, and reprogramming.

DNA methylation is an important regulator of gene transcription and a large body of evidence has demonstrated that aberrant DNA methylation is associated with unscheduled gene silencing, and the genes with high levels of 5-methylcytosine in their promoter region are transcriptionally silent. DNA methylation is essential during embryonic development, and in somatic cells, patterns of DNA methylation are generally transmitted to daughter cells with a high fidelity. Aberrant DNA methylation patterns have been associated with a large number of human malignancies and found in two distinct forms: hypermethylation and hypomethylation compared to normal tissue. Hypermethylation is one of the major epigenetic modifications that repress transcription via promoter region of tumour suppressor genes. Hypermethylation typically occurs at CpG islands in the promoter region and is associated with gene inactivation. Global hypomethylation has also been implicated in the development and progression of cancer through different mechanisms. This chapter will focus on DNA methylation as the major epigenetic mechanism involved in normal biological processes and abnormal events leading to cancer development. It will also focus on the interaction between DNA methylation and other epigenetic mechanisms.

The histone variants of the H2A family are highly conserved in mammals, playing critical roles in regulating many nuclear processes by altering chromatin structure. One of the key H2A variants, H2A.X, marks DNA damage, facilitating the recruitment of DNA repair proteins to restore genomic integrity. Another variant, H2A.Z, plays an important role in both gene activation and repression. A high level of H2A.Z expression is ubiquitously detected in many cancers and is significantly associated with cellular proliferation and genomic instability. This review summarises the current understanding of these variants and their functions, as well as their links to cancer development. Furthermore, the significance of dysfunction of these variants is highlighted with respect to their potential as biomarkers and as new targets for anticancer therapy.

A wealth of nucleobase and ribose modifications have been identified in multiple types of RNA including tRNAs, rRNAs, mRNAs, and small regulatory RNAs. Among them, 5-methylcytosine (m^5C) has been detected in rRNAs, tRNAs, and early reports have indicated its presence in mRNAs. Well established as an epigenetic mark in DNA, the prevalence and function of m^5C in RNA is either incompletely explored (tRNA, rRNA) or virtually unknown (mRNA, other noncoding RNA). Two eukaryotic m^5C RNA methyltransferases have been identified; however, their substrate specificity and biological roles are incompletely understood. With recent advances in bisulfite sequencing of RNA, comprehensive analyses to determine the occurrence and functions of m^5C in the transcriptome now appear feasible. In this chapter, we summarise the current knowledge in this field, focussing primarily on eukaryotic transcriptomes.

Paramutation is a fascinating phenomenon in which epigenetic information can be transmitted through *trans*-interactions between one allele of a gene to another allele or between homologous DNA sequences that establishes a state of gene expression that is heritable for generations. Paramutation was discovered in maize and similar phenomena have been described in other plants, fungi and animals. In this chapter, we describe several classic plant paramutation systems and discuss recent advances that implicate a role for RNA and a number of components of an RNA-based transcriptional silencing pathway on paramutation.

DNMT3L (DNA methyltransferase 3 like) is member of the DNA methyltransferase family of enzymes responsible for the methylation of CpG dinucleotides. Biochemical studies have revealed that while DNMT3L lacks DNA methyltransferase activity, it can bind to and stimulate the activity of *de novo* DNA methyltransferases DNMT3A and DNMT3B. DNMT3L has also been shown to interact directly with chromatin via its plant homeodomain (PHD) like zinc finger domain. Studies in *Dnmt3L*-deficient mice have revealed that DNMT3L is essential for establishing correct methylation patterns at Retro Transposable Elements (RTE), unique loci and parentally imprinted genes in germ cells, and mice without DNMT3L are rendered infertile. Female *Dnmt3L*$^{-/-}$ mice have apparently normal meiosis but in male *Dnmt3L-/-* germ cells there is asynapsis of chromosomes, and "meiotic catastrophe". Dnmt3L was among the first mammalian genes shown to have a paternal effect, where the genotype of the father (*Dnmt3L*$^{+/-}$) affects sex chromosome aneuploidy in adult and embryonic offspring. This chapter will discuss the role of DNMT3L in establishing DNA methylation patterns during gametogenesis, as well as the proven and potential consequences of DNMT3L deficiency to fertility and somatic and germline genetic disease in light of the increasing evidence that epigenetic reprogramming is a dose sensitive and partially stochastic process.

In the past decade, we have become acquainted with an entire new world of fine regulatory control within cells governed by non-coding RNAs. Although we have not completely explored this world, we know from a few well studied examples that not only gene dosage but protein function also, is fine tuned by non-coding RNAs. It appears that proper cell function is largely dependent on non-coding RNAs, many of which were invisible to us until the advent of high throughput sequencing and tiling array technology. The realisation, that transcripts with no open reading frames are biologically active molecules with multiple functions including regulation of the expression of coding RNA, shifts the power from proteins to non-coding RNAs as key modulators of cellular function. In this review, we discuss the various classes of non coding RNAs and the mechanisms employed by them to achieve this status, with a particular focus on their ability to induce epigenetic modifications.

A large variety of methods to measure DNA methylation have been developed and used extensively over the last twenty years. These have been based on selective restriction digestion of methylated DNA, the capture of methylated DNA by methyl binding proteins or antibodies, or bisulphite conversion of DNA. However, all restriction enzyme based methods are dependent on available restriction sites for methylation-specific restriction enzymes and therefore cannot be used to analyse every CpG site in the genome. Although immunoprecipitation methods are not sequence specific, they are unable to provide methylation data at a single-base resolution. Therefore, both of these approaches are limited in their applicability. On the other hand, bisulphite conversion based methods allow methylation studies directed to any CpG site in the genome.

Sodium bisulphite treatment of DNA converts all unmethylated cytosines into uracil while methylated cytosines remain unchanged, thus transferring an epigenetic difference into a measurable genetic difference. A variety of downstream methods such as Polymerase Chain Reaction (PCR), sequencing, Single Nucleotide Polymorphism (SNP) genotyping and mass spectrometry can be coupled with bisulphite conversion. This chapter provides an overview of some of these methods, concentrating on bisulphite sequencing, methylation-specific PCR, pyrosequencing, MassARRAY EpiTYPER and Infinium Human Methylation27 Bead Chip. Considerations for assay selection and detailed protocols for each method are presented in the end of this chapter.

Methylation-sensitive high resolution melting (MS-HRM) is an inexpensive and robust closed tube screening methodology that enables rapid analysis of locus specific DNA methylation for multiple samples. MS-HRM is based on the differential melting behaviour of PCR amplification products derived from methylated and unmethylated templates after bisulfite treatment. The melting profiles of an unknown sample are compared to the melting profiles of standards with known DNA methylation levels. MS-HRM has the advantage of allowing ready distinction between homogenous and heterogeneous DNA methylation. Estimation of DNA methylation to quite low levels in a semi-quantitative manner is possible for homogeneously methylated templates where all the CpG sites are methylated. However for heterogeneous methylation, the formation of multiple heteroduplexes makes quantitation difficult. Digital MS-HRM utilises limiting dilution to amplify single templates enabling DNA methylation analysis at a single allele resolution and enables the quantitative analysis of both homogenous and heterogeneous methylation. The PCR products either generated by MS-HRM or digital MS-HRM can subsequently undergo Sanger DNA sequencing or pyrosequencing to investigate DNA methylation at individual CpG positions.

The location and density of nucleosomes in the eukaryotic genome plays a role in regulating nuclear processes including transcription, replication, recombination, and repair. Microarray mapping of nucleosomally protected DNA has emerged as a powerful, cost-effective, high-throughput method to analyse the relationship between nucleosome position and genomic regulation. In this chapter we discuss experimental considerations such as sample preparation and microarray design. In addition, two procedures are detailed: (1) Formaldehyde Crosslink and Harvest Cells, Isolate Nuclei, and MNase cut chromatin, and (2) Isolation of mononucleosomally protected DNA and fluorescent labelling of DNA for microarray hybridisation. With the information specified in this chapter, most any laboratory equipped for molecular biology with institutional or commercial access to microarray facilities should be should be able to map nucleosome position and occupancy.

DNA methylation plays an essential role in normal human development, where abnormalities in proper establishment and maintenance of DNA methylation patterns result in human disease. Many experimental approaches have been developed for assaying DNA methylation patterns, including enzymatic-based approaches. In this chapter, we highlight some of these approaches and describe their relative advantages and disadvantages. We also describe advances in microarray and sequencing technologies that have improved resolution of enzymatic-based methods, providing expanded coverage of CpG dinucleotides throughout the genome. These approaches are important tools in characterising the role of DNA methylation in genome organisation and function.

With the advances in traditional genetics failing to provide causal genes for many complex diseases, the focus of research is shifting towards determining the importance of gene-environment interactions, more specifically epigenetic regulation. Paramount to answering this question is the knowledge of which transcription factors bind to what sequence, as well as a detailed understanding of how the transcriptional state of a genetic sequence is epigenetically distinguished. The chromatin immunoprecipitation (ChIP) strategy has proven to be a powerful tool to investigate these mechanisms, but has been limited for a long time to single locus analysis. The recent emergence of next-generation sequencing (NGS) technology however has revolutionised the field of epigenomics and for the first time enabled unbiased genome-wide analysis of ChIPed DNA. Here we provide a detailed discussion and protocol on how to best perform ChIP for NGS analysis.

Over the past decades it has become ever more apparent that understanding the genome did not stop with unravelling the genetic code. Regulatory mechanisms are needed to determine which parts of the genome are active or inactive, and to form a memory system that can be passed on over multiple cell divisions for proper functioning of the cell. These mechanisms underpinning heritable gene regulation are encapsulated by the term “epigenetics” and include histone modifications, miRNA and non-coding RNA expression, and methylation of cytosine residues in DNA. In this review, we compare the various methods that can be used to analyse DNA methylation patterns throughout the genome, and discuss their advantages and disadvantages. In addition, we present a detailed protocol for genome-wide DNA methylation analysis based on the capture of methylated DNA using a methyl-CpG binding domain-based (MBD) protein combined with second generation sequencing.

Since biological phenotype and differentiation is regulated partially through CpG DNA methylation, and this mark is relatively easy to measure, genome-wide profiling of methylation landscape is a popular tool in epigenetic research. The burden then falls on bioinformatics to provide normalisation, quality control, and interpretation, while adopting to the vast number of versatile methods to interrogate methylation profiles. While creating a relational report of the results from either of these methods is critical for data to cross over from lab to lab, and from research to diagnostic and translation purposes, the methods are each introducing their unique bias to the data.

Ideally, one would hope all labs would adopt a single method, but since each method offers a unique advantage, such as price, sensitivity, or confidence, one has to overcome the disparity hurdle via bioinformatic tools. Thus identifying the methodological biases of each technique, and the way to compensate for those in the process of generating a universal CpG methylation prediction on a genome region is the ultimate goal we are describing here. Follow up of CpG methylation with other read outs, such as impact on gene expression or coincidence with locations in the genome, where allelic variation is associated with the investigated phenotype, are also key to proper interpretation of the results.

Maize has served as an excellent model for the study of epigenetic gene regulation for the past several decades. The pioneering work of maize geneticists like Barbara McClintock, Alexander Brink, Marcus Rhoades, and others led to the observation of many fascinating phenomena that were later demonstrated to be epigenetically regulated events. Since these observations were made, a great deal of progress has been made in determining the underlying causes of the phenomena, and many of these examples of epigenetic gene regulation are currently being used to elucidate the mechanisms of epigenetic heritability in plants. Today, these phenomena and the mutants that impact them serve as resources for studying how DNA methylation, chromatin structure, and small RNAs act to influence paramutation, gene silencing, and parent-of-origin dependent dosage compensation. The key attributes and potential contributions of each resource are discussed in the context of understanding the mechanisms and significance of epigenetic gene regulation in large, complex genomes.

Biological experiments are rapidly moving into the information age with the explosion of data generated from rapidly evolving technologies. Microarrays can interrogate many thousands to millions of loci within any one sample, while massively parallel sequencing platforms can essentially measure the entire genome. Keeping up with this rapid pace of data accumulation has required the development of online tools for the processing, analysis and annotation of data generated from large numbers of microarrays or next generation sequencing (NGS) experiments. I will cover a selected range of tools available to the biological researcher, from online discussion forums and blogs, to curated databases that store the data and associated annotations. I will then cover viewer tools that enable visualisation the annotations and finally review tools for assay design for follow up validation of NGS and microarray analyses. I will pay particular attention to DNA methylation and provide the reader with an insight into what is available online for the epigenetics researcher aiming to make sense of epigenomic data. Epigenetics can appear as an impenetrable subject; not just to those encountering it for the first time, but to those within the field too. However, epigenetics, like any subject can be made easier to understand using a combination of clear language, creative illustrations and even animations and film clips. This chapter aims to point readers of all experiences towards helpful and easy-to-read resources that educate about epigenetics. It is split into two main sections, the first aimed at a lay audience including teachers and high school students and the second, at graduate and postgraduate students and beyond. Each section contains summaries of published articles and web sites. The chapter ends with a short section on epigenetic societies and research networks and a summary table of resources.

It is intended to provide a sample of some of the best short to medium length reviews on general topics within the field of epigenetics and while we cover a wide variety of themes, we apologise for any areas not covered. We cite the URLs of freely-available articles wherever possible, but many articles will require library access. We also urge readers to contact authors or publishers if they wish to distribute any of the articles for teaching purposes.

2

Marine Sedimentology

The Bivalve Effect

Explaining and understanding life cycles is on many people's minds in spring, and McGill Biologist Dr. Frédéric Guichard is no exception – in fact, he's made a fascinating discovery relating to the life, death, reproduction and communication ... of mussels. Guichard says marine life can communicate over thousands of kilometres, calling into question current fishery management and marine preservation practices. "If I kill mussels in San Diego, it will have an impact in Seattle. We now know that populations are connected," he said. Using mathematical modelling and data from natural populations, Guichard and his colleagues, Dr. Tarik Gouhier and Dr. Bruce A. Menge at Oregon State University, found a phenomenon similar to the "butterfly effect," whereby the actions of one individual can cause a series of chain reactions. Mussel populations communicate by actions such as releasing larvae or dying. "Current practices are based on the knowledge that a mussel can travel no further than 100 km in its lifespan, so efforts are focused on local areas in the belief that we can control local populations," Guichard explains. "But this 'fence approach' only looks at the life history of an animal, which isn't enough to predict how it will affect its environment and other marine life.

"We can now see what we normally don't look for in the wild, so we can use this model to better manage their numbers. Scientists have long theorised about this, but this is the proof," Guichard said.The principles of their discovery should be applicable to many species and will have important ramifications in the short term for the design of marine reserves and in the longer term for fisheries management. Frédéric Guichard was funded by the James

McDonnell Foundation and the research was published in the *Proceedings of the National Academy of Sciences.*

Studies Show Marine Reserves Can be an Effective Tool for Managing Fisheries

Studies conducted in California and elsewhere provide support for the use of marine reserves as a tool for managing fisheries and protecting marine habitats, according to biologists at the University of California, Santa Cruz. A recent study in the Gulf of California, for example, confirmed the validity of a key concept behind marine reservesthe idea that offspring produced in a protected area can replenish the stocks of harvested species outside the reserve.

"It seems really obvious, but it had never been tested," said Peter Raimondi, professor and chair of ecology and evolutionary biology at UCSC and coauthor of a paper describing the findings in the journal *PLoS ONE.* "We created a model to predict the dispersal of larvae outside the reserves, and the results were completely consistent with our predictions," he said.

Raimondi is involved in a collaborative project (called PANGAS) in which researchers are working with Mexican fishing communities to study and manage fisheries in the northern Gulf of California. Local fishermen in the area of Puerto Peñasco set up a network of marine reserves as part of a community-based effort to manage their resources. Ecological and social studies conducted before, during, and after the establishment of those reserves enabled the researchers to track the results. Raimondi emphasized that resource managers have a wide range of tools at their disposal and must take into account both biological and social factors in choosing the best approach. Many species, such as tuna and squid, move around too much to be protected by setting aside certain areas. For species that tend to stay put, marine protected areas can range from no-take reserves to various levels of limited harvesting, and sometimes involve restrictions on who can harvest fish in an area rather than how much can be taken.

The establishment of marine protected areas along the California coast, as called for in the 1999 Marine Life Protection Act (MLPA), has been controversial. A network of protected areas was established on the Central Coast in 2006, and a plan for the North-Central Coast was adopted in August 2009. In Southern California, a task force will soon make recommendations to the state Fish and Game Commission, while on the North Coast the planning process is just getting started.

Raimondi and Mark Carr, also a professor of ecology and evolutionary biology at UCSC, have been actively involved in this initiative. In addition to serving on science advisory teams, they are engaged in an intensive monitoring program to track the effects of the reserves that have already been established.

"We are monitoring those areas at unprecedented levels. It's a comprehensive effort to characterise the populations and the ecosystems so that we can compare the responses to different types of protection," Carr said. "Monitoring studies around the globe systematically show positive responses within protected areas. We want to really identify what aspects of reserve design are important in influencing those benefits."

According to Carr, it will take a few more years of monitoring to see the effects of the Central Coast reserves. In the Channel Islands, however, where reserves were established in 2003 (separately from the MLPA process), surveys have yielded the kinds of results scientists expect to see in protected areas. For example, fish species targeted by fishermen tend to be bigger and more plentiful within the reserves.

This effect is important, because studies have shown that larger, older females are much more important than younger fish in maintaining healthy populations of species such as West Coast rockfish.

"When you have a protected population, you not only get spillover effects when fish swim out of the reserve and get caught, you also have major effects on larval production," Carr said. "The bigger, older fish in the reserve produce a lot of larvae that replenish the fished populations outside."

Carr, who contributed to a report on the first five years of monitoring in the Channel Islands, said that the conclusions are limited by a lack of data collected before the reserves were created. It is possible that some of the observed differences existed before the areas were protected, but such doubts will be erased if current population trends continue, he said.

In Puerto Peñasco, the shellfish harvested by local fishermen grow and reproduce quickly. As a result, the fishermen saw beneficial effects within a year after they had established a network of reserves. Subsequent events, however, underscored the role of social factors in the success of fishery management efforts. A second paper, published in *PLoS ONE* in July, describes how, after its initial success, the local reserve system collapsed due to poaching by outsiders.

"The whole thing got wiped out due to disruption of the social structure that had supported it," Raimondi said. "Scientifically it was really interesting, but for the people who experienced it on the ground, it was terrible."

Richard Cudney-Bueno, a research associate at UCSC's Institute of Marine Sciences and cofounder of the PANGAS project, is the lead author of both papers. "Here was a group of fishermen that had already seen some declines in the shellfish they harvested. This led to the implementation of community-based efforts to manage their resources, including the establishment of marine reserves," he said. "We found that local control of community resources can work, but there has to be broader government support to back up the local efforts."

A native of Mexico City, Cudney-Bueno has been working with Mexican fishing communities and conducting ecological and social research in the Gulf of California since the mid 1990s. He now has a joint position with UCSC, the University of Arizona, and the David and Lucile Packard Foundation.

The first reserve in the Puerto Peñasco area was established in 2001 around an island. Cudney Bueno and other researchers, working with a Mexican nonprofit organisation (Centro Intercultural de Estudios de Desiertos y Océanos), trained the fishermen to monitor shellfish populations in and around the reserve. "The response was really quick, so they could see a classic reserve effect one year later," Cudney-Bueno said. "That led to more areas being closed, and the first paper shows the effects of the network of reserves."

The cooperative was so successful it was recognised by the Mexican government with a Presidential Conservation Award. But word spread quickly along the coast about the thriving shellfish populations in Puerto Peñasco, and other fishermen from outside the community began to move in and poach from the reserves. After poaching began, the system of cooperation that had established and protected the reserves broke down.

Now, the situation is beginning to improve again, Cudney-Bueno said. The Mexican government has created one of a handful of exclusive fishing zones in the Gulf of California, giving the local cooperative the exclusive legal right to harvest shellfish in the Puerto Peñasco area.

"They now have a strong management plan with legal rights and government support, so I think they will be able to get back to where they were before the poaching started," Cudney-Bueno said. "I see it as part of the evolution of a management system. Social change takes time, and it really hasn't been that long. A lot is happening now in Mexico and around the world as local people are increasingly asking for control over their resources. Various fishing communities in Mexico, including lobster and abalone fishermen in Baja California, have moved forward with the establishment of their own marine reserves and government-backed forms of territorial-use rights."

The which brings together experts from UCSC, the University of Arizona, and several collaborating academic institutions and nonprofit organisations in Mexico, is working with other fishing communities in the Gulf of California to develop management plans for the region's marine resources.

"PANGAS is now working with the Mexican government to build management plans for a series of species in the northern Gulf of California," Raimondi said. "It's interesting to compare that with the MLPA process in California. The approaches are very different, and it has to do with differences in government and social structures."

According to Carr, the California MLPA process is now being used as a model in other parts of the world, most notably in the United Kingdom. Additional information about the MLPA process can be found on the California Department of Fish and Game web site at

Marine Reserves Hit the Spotlight in PNAS Special Issue, AAAS Press Briefing

Marine reserves are known to be effective conservation tools when they are placed and designed properly. This week, a special issue of the *Proceedings of the National Academy of Sciences* (PNAS) is dedicated to the latest science on marine reserves, with a focus on where and how reserves can most effectively help to meet both conservation and fisheries goals. "There is plenty of new evidence to show that if reserves are designed well, they can benefit both fish and fishermen," explains Steven Gaines, newly appointed Dean of the Bren School of Environmental Science & Management at UC Santa Barbara and a guest editor of the PNAS special issue. "An enormous amount of research has already been done on marine reserves, helping to facilitate their use and development around the world, and yet many lingering questions remain. Papers in this special issue help answer many of those questions."

Several scientists with papers in the issue will be discussing their findings today (Sunday, February 21, 2010, at 3 p.m. U.S. Pacific time)

at a press briefing at the Annual Meeting of the American Association for the Advancement of Science (AAAS) in San Diego.

Each paper in the special issue describes a different yet interrelated aspect of reserves and the societies in which they function. One global study, for example, identifies large stretches of coastal ocean where marine reserves can play a major role in reducing cumulative impacts on marine ecosystems. In some cases, fishing is responsible for as much as 90 percent of the overall damage to the ecosystem. These areas include the South China Sea, East China Sea, Bering Sea, Sea of Okhotšk, North Sea, and Norwegian Sea, as well as much of the Coral Triangle. When fishing is the most significant threat to a particular ecosystem, marine reserves can serve as powerful tools for improving the overall condition of the ocean.

"It was surprising to see how much marine reserves could improve overall ocean health in many places around the world, but also humbling how much still needs to be done," says Ben Halpern, an associate research biologist at UCSB and lead author of this study. "Marine reserves and other marine protected areas are an important piece of the puzzle in addressing marine resource management comprehensively, but they are only part of the solution."

Another global study – the first of its kind – examined the social determinants of success in 56 coral reef marine reserves, representing every major reef region in the world. The results suggest a clear link between reserve performance and proximity to a human population centre. However, the overall effect of having humans nearby – beneficial or detrimental – varied by region. In the Indian Ocean, for example, reserves near a human population performed better than their more remote counterparts. In the Caribbean, however, reserves near a population centre fared relatively poorly.

"Success also has to do with cooperation by the local communities," explains Joshua Cinner, Senior Research Fellow at James Cook University in Australia and a co-author on this study. "In areas where people work together to invest in their resources, we saw less poaching inside marine reserves. This was really about having processes that allow people to be consulted about the reserves and engage in research and management activities. Our study found that, to get high levels of compliance with reserve rules, managers need to foster the conditions that enable participation in reserve activities, rather than just focusing on patrols."Detailed spatial information on where fish spawn and spend their lives can help in the design of effective reserve networks.

To that end, another study looks at what ecologists call "sources" and "sinks." Source areas are very productive spawning grounds that are well connected to other areas by ocean currents. Larvae hatched there are swept to the sink areas, where they seed populations that might otherwise dwindle to unsustainable levels. The study suggests that, if fishing is prevented in source areas and instead concentrated in sink areas, fisheries could realise a significant gain in value – greater than 10 percent according to model simulations.

"If you care about both conservation and fisheries, regardless of the relative value you place on each, better information about dispersal will help you achieve both goals," explains Andrew Rassweiler, a biologist at UC Santa Barbara and a co-author on this study. "Because managers have to be selective when placing reserves, information on sources and sinks will help them get the best results with the least disruption to fishing."

"What you'd really like to do is close the source to fishing and only fish in the sink area," adds Christopher Costello, an economist at the Bren School and lead author on this paper. "It turns out you get a much higher economic value and much better conservation when you do that. But if you don't know where the sources and sinks are, you can't do that, so that is where the information comes in."

Finally, another study presents findings from the Channel Islands National Marine Sanctuary, home to a network of marine protected areas first established by the California Department of Fish and Game in 2003. The authors found that, after only five years, fished species were significantly larger and more abundant inside no-take reserves than they were outside.

"We're at the point where we can actually begin to assess network benefits, including increases in the size and number of fish across the entire network," says Jennifer Caselle, a research biologist at the Marine Science Institute at UC Santa Barbara. "Based on what we know so far, it seems the whole really can be more than the sum of its parts when networks are designed with larval and adult movements in mind."

Many scientists point to networks as a key strategy for getting the most out of reserves. In fact, the special issue and press conference coincide roughly with the implementation of a comprehensive network of marine reserves in Southern California mandated by the California Marine Life Protection Act (MLPA). The MLPA network (of which Southern California represents the third of five stages) is the first in the United States to be designed from the ground up as a network rather than a patchwork of independent reserves.

The scientists emphasize that the coupling of new marine reserve network science with appropriate public process is the key.

"If the public and stakeholder process and networks are well designed and implemented, it will work, even in an area like California with tens of millions of users," says Gaines. "These networks can benefit both fish and fishermen. It's not a choice."

Scientists Find Community Involvement, Not only Enforcement, Drives Success of Marine Reserves

In one of the most comprehensive global studies of marine reserves, a team of natural and social scientists from the University of Rhode Island and other institutions has found that community involvement is among the most important factors driving the success of marine reserves. "We make a big mistake thinking that a marine reserve is just about coral, fish and other aquatic organisms," said Richard Pollnac, URI professor of anthropology and marine affairs, who led the study. "They are also composed of the people who can make them succeed or fail and who are either helped or hurt by them."

The study was published in the *Proceedings of the National Academy of Sciences* on February 22.

The researchers studied 127 marine protected areas in the Caribbean, Western Indian Ocean and the Philippines to identify the key factors that determine the success of marine reserves, which protect the marine environment by prohibiting fishing. Biological assessments were conducted at 56 of the reserves to determine their ecological health, while surveys of residents and community leaders in local communities discerned perceptions and opinions in all 127 reserves.

Among their results, the researchers found that the reserves where residents said they complied with the rules were more effective at protecting fish stocks than those where the rules were often ignored. They noted, however, that compliance with reserve rules occurred not only due to surveillance and enforcement but also due to complex social interactions among community members and opinion leaders.

"The most successful reserves were those where the people said that most of the community follows the rules," explained Graham Forrester, URI associate professor of natural resources science and a co-author of the study. "Compliance with the rules is a measure of how a community feels about the reserve. It's their choice to follow the rules."

The researchers noted that their surveys indicated that it is vital to the success of any marine reserve that community members are participants in the process of setting up and monitoring the reserve.

Other research results were somewhat surprising.

The effect of human population density near marine reserves, for instance, differed significantly from location to location. As the researchers expected, greater population density negatively impacted reserves in the Caribbean, but it had no detectable affect at marine reserves in the Philippines. At reserves in the Western Indian Ocean, on the other hand, greater population density was correlated with healthier reserves and greater fish biomass inside the reserve compared with outside. Study co-author Tracey Dalton, URI associate professor of marine affairs, said that it is not easy to explain these disparities. The positive effects in the Indian Ocean may be driven by increased fishing pressure outside the reserve or the result of people migrating to areas where the marine reserves are most successful.

"It's important to recognise that people are part of the ecology of marine reserves," Pollnac said. "If you can demonstrate to them that the reserve will have more fish while also providing benefits to the community, and if you pay attention to the needs of the people, then there's a much greater chance that the reserve will be a success."

Participaton 'Important for Healthy Marine Parks

he involvement of locals is a key ingredient in the success of marine parks which protect coral reefs and fish stocks. The largest-scale study to date of how coastal communities influence successful outcomes in marine reserves has found that human population pressure was a critical factor in whether or not a reserve succeeded in protecting marine resources – but so too was local involvement in research and management.

The team looked at how successful coral reef marine reserves were at conserving fish stocks. They studied 56 marine reserves from 19 different countries throughout Asia, the Indian Ocean, and the Caribbean.

"About ¾ of the marine reserves we studied showed a positive difference in the amount of fish inside compared to outside – so most reserves we studied were working" says Dr Josh Cinner of the ARC Centre of Excellence for Coral Reef Studies and James Cook University

"However, the differences weren't always large. The most successful reserves showed really big differences of 14 times the amount of fish inside compared to outside, but that wasn't always the case.

"What we were most interested in, was understanding what made some reserves more successful than others. One of the best predictors of how 'successful' a marine reserve was, is actually the size of the human communities around the reserve – but interestingly, this varied in different regions.

"In the Indian Ocean, for example, where reserves are government-controlled and moderate in size (around six square kilometres on average), having lots of people nearby had a positive effect. But this could be because marine resources outside the reserve are heavily degraded, accentuating the healthier state of those inside the reserve.

"In the Caribbean, we found the opposite. Large human populations near reserves led to poor performance of the reserve – which may be due to low compliance or poor enforcement in marine parks near population centres," Dr Cinner said.

The other key ingredient for a successful marine reserve was the level of poaching in the reserve. But importantly, the team found that compliance with reserve rules was not just related to the level of enforcement, but also to a range of social, political, and economic factors which enabled people to co-operate better in protecting their marine resources. Reserves worked best where there was a formal consultation processes about reserve rules, where local people were able to participate in monitoring the reserve, and when ongoing training for community members was provided so that they could better understand the science and policy.

"It was clear that this type of local involvement was a very important factor in building the local support necessary to make reserves successful. Park agencies need to foster conditions that enable people to work together to protect their local environment, voluntarily, rather than focusing purely on regulations and patrols.

"Enforcement will almost always be an important part of a successful reserve, but there is a lot of ocean out there to patrol and many of the places we studied were poor, developing countries which don't have the luxury of being able to invest in lots of patrol boats.

The team's report appears in the *Proceedings of the National Academy of Sciences* (PNAS) in a paper entitled "Marine reserves as linked social–ecological systems" by Richard Pollnac, Patrick Christie, Joshua E. Cinner, Tracey Dalton, Tim M. Daw, Graham E. Forrester, Nicholas A. J. Graham, and Timothy R. McClanahan.

New Study Shows Marine 'Networks' Can Protect Fish Stocks

University of Miami (UM) Rosenstiel School of Marine & Atmospheric Science faculty were part of an international scientific team to show that strong links between the corals reefs of the South China Sea, West Pacific and Coral Triangle hold the key to preserving fish and marine resources in the Asia-Pacific region.

Rosenstiel School researchers Drs. Claire Paris and Robert Cowen and colleagues from the ARC Centre of Excellence for Coral Reef Studies at James Cook University and University of California - Los Angeles, have established that the richest marine region on Earth – the Coral Triangle between Indonesia, Malaysia and the Philippines – depends vitally for its diversity and resilience on coral and fish larvae swept in from the South China Sea and Solomon Islands.

"By evaluating the directionality of larval transport over multiple generations, we could describe the signature of the extraordinary genetic diversity of the Coral Triangle. Preserving diversity is key to the health of marine systems," said Claire Paris, Rosenstiel School assistant professor of Applied Marine Physics. "This kind of work will help us anticipate and manage changes of connectivity networks in the future."

The authors provide evidence showing the regions' biology is closely inter-connected suggesting that it is in the interests of all Asia-Pacific littoral countries to work together more closely to protect it.

"Maintaining the network of links between reefs allowing larvae to flow between them and re-stock depleted areas, is key to saving coral ecosystems threatened by human pressure and climate change," said the paper's lead author Johnathan T. Kool of James Cook University, who is also an alumnus of UM. "The science shows the region's natural resources are closely interconnected. Nations need to cooperate to look after them and that begins with recognising the resources are at risk and that collective action is needed to protect them."

The Coral Triangle is home to more than one third of all the world's coral reefs, including over 600 different species of reef-building coral and 3,000 species of reef fish. These coral ecosystems provide food and income for more than 100 million people working in marine-based industries throughout the region. Six nations within the Coral Triangle – Indonesia, the Philippines, Malaysia, Papua New Guinea, The Solomon Islands and Timor L'Este are now working together to strengthen coral reef governance and management, under an arrangement known as the Coral Triangle Initiative.

Coral 'Network' Can Protect Asia-Pac Fish Stocks

An international scientific team has shown that strong links between the corals reefs of the south China sea, West Pacific and Coral Triangle hold the key to preserving fish and marine resources in the Asia-Pacific region. Research by Dr Johnathan Kool of the ARC Centre of Excellence for Coral Reef Studies and James Cook University, and his colleagues, has established that the richest marine region on Earth – the Coral Triangle between Indonesia, Malaysia and the Philippines – depends vitally for its diversity and resilience on coral and fish larvae swept in from the South China Sea and Solomon Islands.

"The currents go in various directions, but the prevailing direction is from east to west, and this carries coral spawn and fish larvae from areas such as round the Spratly Islands in the South China Sea and the Solomons/Papua New Guinea," he explains.

"Maintaining the network of links between reefs allowing larvae to flow between them and re-stock depleted areas, is key to saving coral ecosystems threatened by human pressure and climate change.

"The Coral Triangle is home to more than one third of all the world's coral reefs, including over 600 different species of reef-building coral and 3,000 species of reef fish. These coral ecosystems provide food and income for more than 100 million people working in marine based industries throughout the region," Dr Kool explains.

"Knowing where coral spawn comes from is vital to managing our reefs successfully. Even though coral reef communities may not be connected directly to one another, reefs on the edge of the Coral Triangle have the potential to contribute significant amounts of genetic diversity throughout the region," says Dr Kool.

He argues that recent evidence showing the region's biology is closely interconnected suggests it is in the interests of all Asia-Pacific littoral countries to work together more closely to protect it: "The science shows the region's natural resources are closely interconnected. Nations need to co-operate to look after them and that begins with recognising the resources are at risk and that collective action is needed to protect them. Six nations within the Coral Triangle, (Indonesia, the Philippines, Malaysia, Papua New Guinea, The Solomon Islands and Timor L'Este) are now working together to strengthen coral reef governance and management.

How Corals Fight Back?

Australian researchers are a step closer to understanding the rapid decline of our coral reefs, thanks to a breakthrough study linking coral immunity with its susceptibility to bleaching and disease. The discovery was made by Caroline Palmer, Bette Willis and John Bythell, scientists from the Australian Research Council Centre of Excellence for Coral Reef Studies, James Cook University (Queensland) and Newcastle University (UK).

"Understanding the immune system of reef-building corals will help to reduce the impact of coral diseases and environmental stresses," says Caroline Palmer, lead author of the publication. "Potentially, this will enable us to more accurately predict the vulnerability of coral reefs to disease and bleaching, before there are obvious signs of stress."

"This unique study broadens the limited knowledge we have about the defence systems of corals, which is one of the main challenges facing scientists aiming to protect corals" says Professor Bette Willis, a chief investigator in the ARC CoE for Coral Reef Studies. "Identifying and measuring the immune functions of several different corals allows us to predict which ones are particularly susceptible to stress."

"Variation in levels of immune function among different species is likely dependent on the energy they assign to it. As energy is vital for an effective immune response, corals that utilise energy to grow and reproduce rapidly have less to spare for their immune response," says Caroline Palmer. "These corals, like the staghorns, Acropora, are the colonies most vulnerable when challenged by temperature stress or disease."

A key element of the coral immune system is melanin production. Melanin, a classic part of immune responses found in invertebrates, also provides a defence against disease-causing organisms in corals. It may also be used to stop harmful UV light from reaching the symbiotic algae and causing bleaching.

The study of coral immunity will enable scientists to better pre-empt the effects of different stresses on corals. This is important, as by the time physical symptoms become apparent, strategies to mitigate stress effects will be far less valuable. "Our increased understanding of coral immune systems may therefore be used to address the causes rather than the symptoms of coral declines", says Caroline Palmer. Bette Willis adds: "This approach is necessary particularly given that coral bleaching is similar to having a fever – it's a common sign for many different stresses so it's often difficult to point to any one cause in particular".

Two of the main factors that cause corals to bleach are attacks by disease-causing microbes and temperature stress. It is currently estimated that between three and six per cent of corals in the Great Barrier Reef (GBR) are affected by coral diseases, and up to a third of corals at a given location can be affected by temperature stress in a warm year. Temperature stress is a growing concern due to global warming.

Researchers and reef managers are currently working on strategies to protect vulnerable coral sites. The preventive measures envisioned involve minimising human impacts which might further injure the coral, such as dredging, building construction, pollution, land runoff or damaging corals by boat activity or fishing.

Coral Bleaching Increases Chances of Coral Disease

Mass coral bleaching has devastated coral colonies around the world for almost three decades. Now scientists have found that bleaching can make corals more susceptible to disease and, in turn, coral disease can exacerbate the negative effects of bleaching. A paper in the October issue of the journal *Ecology* shows that when they occur together, this combination of afflictions causes greater harm to corals than either does on its own. "Traditionally, scientists have attributed coral declines after mass bleaching events to the bleaching only," says Marilyn Brandt, a post-doctoral researcher at the University of Miami and the lead author on the paper. "This study shows that the interplay between diseases and bleaching can play a much larger role than we realised."

Corals rely on algae that live inside each coral polyp to provide nutrients and supplemental oxygen. Bleaching occurs when these colourful algae die out or leave the polyps, often in response to overly warm conditions. Without their brightly coloured algae, the coral's skeleton becomes visible through its transparent tissue, making it appear white. Although the tissue remains intact and can recover over time, this stressful condition can cause corals to stop growing and reproducing. Warmer water temperatures can also lead to increased incidence of coral diseases, which, unlike most bleaching, can cause irreparable loss of coral tissues. In many cases, bleaching and disease occur concurrently on coral reefs. Brandt and her colleagues wondered if the occurrences of bleaching and disease were linked beyond simply occurring under the same conditions.

"Coral bleaching and coral diseases are both related to prolonged thermal stress," says Brandt. "But we wanted to look closer to find out whether they were interacting and what was actually causing the decline we see."

In the summer and fall of 2005, the same oceanic temperature shifts that contributed to the creation of Hurricane Katrina caused a warm mass of water to settle over the northeast Caribbean and parts of Florida. This sustained warming triggered a mass coral bleaching event that affected up to 90 percent of coral reef cover in the area. Brandt and her colleagues surveyed colonies in the Florida Keys before, during and after this event to determine the relationship between bleaching and coral disease.

The researchers found that the coral diseases they observed were related to bleaching, but in different ways. The prevalence of white plague disease increased during the bleaching event, an observation that Brandt says may have to do with increased susceptibility to the disease.

"Higher temperatures can increase the growth rate of coral pathogens, such as bacteria and other microbes, so we probably see a higher disease incidence because of the expansion of these pathogens in the environment," Brandt says. "But bleaching is also a stressed state, and just like any other animal under stress, the coral's disease resistance is lowered."

In addition, the researchers found that colonies already infected with another disease, known as dark spot disease, suffered more extensive bleaching than healthy corals. Brandt thinks a fungus that's likely associated with this infection could cause the relationship of the algae and the coral to be weakened, leaving the corals more susceptible to bleaching.

Because diseases happen on a much finer scale than mass bleaching events, Brandt says that more informed management of coral ecosystems should involve more frequent monitoring to determine the underlying causes of coral damage. "Understanding how these different stressors interact can help explain the mortality pattern we see after large-scale bleaching events," says Brandt. "If we understand what's causing the mortality, we can institute control measures that are more specific to the causes.

Modest Fisheries Reduction could Protect Vast Coastal Ecosystems

A reduction of as little as five per cent in fisheries catch could result in as much as 30 per cent of the British Columbia coastal ecosystems being protected from overfishing, according to a new study from the UBC Fisheries Centre. The study, by Natalie Ban and Amanda Vincent of Project Seahorse, proposes modest reductions in areas where fisheries take place, rather than the current system of marine protected areas which only safeguard several commercially significant species, such as rockfish, shrimp, crab, or sea cucumber. The article is published today in *PLoS ONE*, an online journal of the Public Library of Science.

Using B.C.'s coastal waters as a test case, the study affirms that small cuts in fishing, if they happen in the right places could result in very large unfished areas. For example, a two per cent cut could result in unfished areas covering 20 per cent of the B.C. coast, offered real conservation gains.

"The threat of over-fishing to our marine ecosystems is well-documented," says Ban, who recently completed her PhD at the UBC Fisheries Centre. "Our study suggests a different approach could reduce the impacts on fishers as well as helping us move towards achieving conservation goals."

Part of the reason for the research was to open a debate on how to meet conservation goals set during the 2002 World Summit on Sustainable Development, which included establishing a network of marine protected areas by 2012.

"With the current rates of progress, there is no chance of meeting our 2012 targets," says Ban. "Given that fishers recognise the problem of overfishing but often regard marine protected areas as serving only to constrain them, another approach must be found. That's why we undertook this study."

The research looked at spatial catch data from Fisheries and Ocean Canada for 13 commercial fisheries on Canada's west coast to show that large areas representing diverse ecoregions and habitats might be protected at a small cost to fisheries. "Given the dismal state of many fisheries, we urgently need to identify alternative approaches to sustaining marine life while respecting the needs of fishers and fishing communities," says Amanda Vincent, Canada Research Chair in Marine Conservation at UBC and Project Seahorse director. "We have little to lose – and much to gain – in trying a new approach in areas where marine conservation remains inadequate. Our research is globally relevant

A Global Model for the Origin of Species Independent of Geographical Isolation

The tremendous diversity of life continues to puzzle scientists, long after the 200 years since Charles Darwin's birth. However, in recent years, consistent patterns of biodiversity have been identified over space, time organism type and geographical region. Two views of the process of "speciation" the evolutionary process by which new biological species arise dominates evolutionary theory. The first requires a physical barrier such as a glacier, mountain or body of water to separate organisms enabling groups to diverge until they become separate species. In the second, an environment favours specific characteristics within a species, which encourages divergence as members fill different roles in an ecosystem.

In a new study, "Global patterns of speciation and diversity," just published in *Nature*, Les Kaufman, Boston University professor of biology and associate director of the BU Marine Program along with a team of researchers from The New England Complex Systems Institute, have collaborated and found a way to settle the debate which deals with the origin of species independent of geographic isolation.

They demonstrated, using a computer model, how diverse species can arise from the arrangement of organisms across an area, without any influence from geographical barriers or even natural selection. Over generations, the genetic distance between organisms in different regions increases, the study noted. Organisms spontaneously form groups that can no longer mate resulting in a patchwork of species across the area. Thus the number of species increases rapidly until it reaches a relatively steady state.

"Our biodiversity results provide additional evidence that species diversity arises without specific physical barriers," the study states.

The computer simulations, the authors, note showed the distribution of species formed patterns similar to those that have occurred with real organisms all around the world.

"The model we put forward in the paper lays the groundwork for more powerful tests of the role played by natural and sexual selection, as well as habitat complexity in shaping the patterns of biological diversity that we see around us today," said Kaufman. Our insights can be applied to the immense challenge that we now face not only to prevent the extinction of a large chunk of life, but also to prevent ourselves from quenching the very forces that fuel the continuous creation of new life forms on earth."

This study is also the fourth in a series from The New England Complex Systems Institute on the role of complexity in species coexistence and evolutionary diversification.

"One can think about the creation of species on the genetic level in the same way we think about the appearance of many patterns, including traffic jams," said Yaneer Bar-Yam, president of The New England Complex Systems Institute and a senior author of the study. "While the spatial environment may vary, specific physical barriers aren't necessary. Just as traffic jams can form from the flow of traffic itself without an accident, the formation of many species can occur as generations evolve across the organisms' spatial habitat.

Composites for Energy

Advanced composite materials are playing a vital role in improved design and reduced operating costs for renewable energy technologies. Research presented today [Tuesday 30 June] will highlight how wind, marine and solar power could address these challenges within the renewable energy industry. The 'Composites for Energy' seminar has been organised jointly between Bristol University's Advanced Composites Centre for Innovation and Science (ACCIS) and the University's BRITE Futures Institute, a new multidisciplinary research hub dedicated to environmental systems and technologies. The seminar is part of this year's annual ACCIS conference.

Recent studies have suggested that around 5 per cent of the UK's electricity needs could be supplied by tidal stream devices and unlike tidal barrages, they do not block tidal channels and have none of the associated environmental impacts such as loss of intertidal habitats for bird and marine life. Tidal stream devices capture energy from fast-flowing tidal currents, such as those found in constrained channels around headlands.

A collaborative project, 'New Materials and Methods for Energy Efficient Tidal Turbines (NEW-MMEETT),' involving the University,

Aviation Enterprises Ltd (blade manufacturer), Advanced Composites Group Ltd (composites manufacturer) and Materials Engineering Research Laboratory Ltd is developing more fatigue resistant materials and improved design techniques for tidal turbine blades. This will enable a reduction in the mass of material required for blade manufacture, essential for making such devices more commercially viable, whilst also ensuring required in-service lifetimes can be met with minimum maintenance requirements.

Bristol's role within the project involves the development of a numerical modelling technique for predicting how damage grows in composite materials under cyclic loading.

Dr Stephen Hallett in ACCIS said: "The outcomes of the NEW-MMEETT project, with respect to both more fatigue resistant materials and improved design techniques, will also have strong potential for application to wind turbine blades." Wind turbines are already a well-established technology but further design and manufacturing improvements are essential in helping the industry meet its expected growth targets over the coming years. The Global Wind Energy Council (GWEC) has predicted that wind energy could provide as much as 13 per cent of global electricity demand in 2020 and as much as 25 per cent in 2030.

Modern wind turbine blades are generally made from a combination of glass and carbon fibre reinforced plastics. During manufacture, the plastic resin is heated and cooled in a controlled manner so that it bonds with the fibres and sets to form a rigid structure. The combination of very strong fibres surrounded by a lightweight plastic matrix enables a greater strength to weight ratio than is possible with conventional metallic materials. By carefully controlling the direction and tension of the fibres, it is also possible to create a bistable composite, which can snap between two distinct rigid shapes.

Dr Weaver in ACCIS said: 'We are currently focused on producing morphing blades, which can rapidly change their aerodynamic profile to best suit the current wind conditions. This has the potential to significantly relieve unwanted stresses in the blades, increasing their efficiency and helping to prolong their life. In addition to wind turbine and helicopter rotor blades, morphing composites are also being developed for aircraft wings, reducing the need for mechanically operated control surfaces.' The Lithiated Nanoparticle Diamond Solar Energy Converter project, based at the University and funded by the energy company, E.ON, plans to exploit solar heat to produce electricity. The project uses a novel application of commercially available, low cost diamond powder to form Lithium-doped nano-Diamond (LiD) electron emitters. The LiD emitters use solar infrared radiation to produce thermionic emission in a vacuum valve. The current and voltage produced by the valve is converted to electrical power that may be fed into the national grid. Such a device is termed a thermionic converter and has the potential to realise theoretical conversion efficiencies of 66 per cent.

Dr Neil Fox, in the University's School of Chemistry, said: "We aim to demonstrate a working nanodiamond based solar energy converter as an alternative technology to conventional photovoltaic solar cells. The target is to achieve operation below red heat. If this can be realised, the new thermionic converter technology will have applications in renewable energy generation particularly for Concentrated Solar Power.

Discovery of the Jekyll-and-Hyde Factors in 'Coral Bleaching'

Scientists are reporting the first identification of substances involved in the Jekyll-and-Hyde transformation that changes harmless marine bacteria into killers that cause "coral bleaching." Their study appears in ACS' *Environmental Science & Technology*, a semi-monthly journal. Dan Bearden and colleagues note that bleaching already has destroyed up to 30 percent of the world's coral reefs, and scientists are searching for ways to slow or stop the damage. One known culprit is an ocean-dwelling bacterium, Vibrio coralliilyticus (V. coralliilyticus) that chokes-off corals' energy supply and kills these shell-clad marine animals. At lower temperatures, the bacteria are harmless to coral. But at warmer temperatures (above 75 degrees Fahrenheit) the bacteria become virulent and can kill coral.

The new study reports identification of three chemicals betaine, glutamate, and succinate that V. coralliilyticus produces in warmer water and are involved in the transformation. The discovery opens the door to understanding the biology involved in the complex interactions between corals and bacteria and unravelling the mystery of coral bleaching, the scientists indicated.

The White Stuff: Marine Lab Team Seeks to Understand Coral Bleaching

With technology similar to that used by physicians to perform magnetic resonance imaging (MRI) scans, researchers from six institutions—including the National Institute of Standards and Technology (NIST)—working at the Hollings Marine Laboratory (HML) in Charleston, S.C., are studying the metabolic activity of a pathogen shown to cause coral bleaching, a serious threat to undersea reef ecosystems worldwide. Coral bleaching is the whitening of living coral due to a disruption of the symbiosis (two organisms whose living together benefits both) with its zooxanthellae, tiny photosynthesizing algae. These unicellular creatures reside within the coral's tissues and provide the host organism with up to 90 percent of its energy. It's the solar-derived chemical products of these algae that give the world's coral species a rainbow of vivid colours. Unfortunately, ecologically valuable coral colonies around the globe are being threatened by an ocean-dwelling bacterium known as *Vibrio coralliilyticus*. When the microbe becomes virulent, it can infiltrate coral and dislodge the zooxanthellae, causing the coral to lose its pigmentation. If symbiosis is disrupted long enough, the coral dies from starvation.

Environmental scientists have shown in laboratory experiments that the virulence of *V. coralliilyticus* is temperature dependent, causing bleaching at temperatures above 24 degrees Celsius (75 degrees Fahrenheit). These findings have raised concerns that increasing ocean temperatures either through natural seasonal changes or climate change trends may lead to increased risk of widespread coral bleaching. During the past two decades, it has been reported that nearly 30 percent of the world's coral reefs and the ecosystems they support have been severely degraded by bleaching.

In a recent paper in *Environmental Science and Technology*,* the HML research team described how it used nuclear magnetic resonance (NMR) to study metabolic changes in V. coralliilyticus resulting from temperature effects. The technique allows discovery of small-molecule metabolism-related compounds that correlate with different biological conditions. In this study, the levels of three compounds- betaine, glutamate and succinate that help regulate energy production and osmotic pressure (a mechanism for maintaining cellular integrity) in *V. coralliilyticus* were determined to vary significantly between 24 degrees Celsius when the bacterium is not virulent and 27 degrees Celsius (81 degrees Fahrenheit) when it is. These metabolic changes, the HML team believes, are clues to learning why the small temperature change can turn non-virulent *V. coralliilyticus* into a coral bleaching menace. Future metabolomic studies of *V. coralliilyticus* are planned to better understand the complete temperature-dependent mechanism involved in its pathogenicity. The researchers hope that these findings will lead to a better understanding of the symbiotic relationships that exist in healthy coral and the potential impacts on those relationships under changing ecological conditions.

Long-term Recovery of Reefs from Bleaching Requires Local Action to Increase Resilience

In the journal *Estuarine, Coastal and Shelf Science,* University of Miami Rosenstiel School of Marine and Atmospheric Science Professor Dr. Peter Glynn, and 2008 Pew Fellow for Marine Conservation and Assistant Professor Dr. Andrew Baker, assess more than 25 years of data on reef ecosystems recovery from climate change-related episodes of coral bleaching. Coral bleaching, in which corals expel their symbiotic algal partners and turn pale or white is one of the most visible impacts of climate change on marine ecosystems. Typically caused by higher-than-normal ocean temperatures, it can lead to widespread death of corals and is a major contributor to the rapid decline of coral reef ecosystems worldwide.

The paper, co-authored by Dr. Bernhard Riegl, associate director of the National Coral Reef Institute, represents the first comprehensive review of long-term global patterns in reef recovery following bleaching events. Bringing together the results of dozens of bleaching studies, the article reports that bleaching episodes set the stage for diverse secondary impacts on reef health, including coral disease, the breakdown of reef framework, and the loss of critical habitat for reef fishes and other important marine animals.

“Bleaching has resulted in catastrophic loss of coral cover in some locations, and has changed the coral community structure in many others,” said Glynn. “These dramatic fluctuations have critical impacts on the maintenance of biodiversity in the marine tropics, which is essential to the survival of many tropical and sub-tropical economies.”

However, the paper also shows that, while bleaching episodes have resulted in dramatic loss of coral cover in certain locations, reefs vary dramatically in their ability to bounce back from these disturbances. It also evaluates factors explaining why some species of coral recover better than others, as well as why some reef regions are recovering while others are not.

The study finds that reefs in the Indian Ocean are recovering relatively well from a single devastating bleaching event in 1998. In contrast, western Atlantic (Caribbean) reefs have generally failed to recover from multiple smaller bleaching events and a diverse set of chronic additional stressors such as diseases, overfishing and nutrient pollution. No clear trends were found in the eastern Pacific, the central-southern-western Pacific or the Arabian Gulf, where some reefs are recovering and others are not.

"These findings illustrate how coral reefs, under the right conditions, can demonstrate resilience and recover from bleaching, even when it initially appears catastrophic", said Baker. "What prevents them from doing so is the lethal prescription of combined, additional stressors that prevent them from recovering in-between recurrent bleaching events. If we can remove or reduce these stressors we might give reefs a fighting chance of surviving climate change".

The paper entitled "Climate change and coral reef bleaching: An ecological assessment of long-term impacts, recovery trends and future outlook" also discusses potential mitigation and intervention strategies that might maximise coral reef survival in the coming years. It concludes that bleaching disturbances are likely to become a chronic stress in many reef areas in the coming decades, and coral communities, if they cannot recover quickly enough, are likely to be reduced to their most hardy or adaptable constituents. Unless significant reductions in greenhouse gas emissions can be achieved within the next two to three decades, maximising coral survivorship during this time may be critical to ensuring healthy reefs can recover in the long term.

Discovery of How Coral Reefs Adapt to Global Warming could Aid Reef Restoration

Discoveries about tropical coral reefs, to be published on 23 June 2010, are expected to be invaluable in efforts to restore the corals, which are succumbing to bleaching and other diseases at an unprecedented rate as ocean temperatures rise worldwide. The research gives new insights into how the scientists can help to preserve or restore the coral reefs that protect coastlines, foster tourism, and nurture many species of fish. The research, which will be published in the journal *PLoS One*, was accomplished by an international team whose leaders include Iliana Baums, an assistant professor of biology at Penn State University. The team focused on one of the most abundant reef-building species in the Caribbean, *Montastraea faveolata*, known as the mountainous star coral. Though widespread, this species is listed as endangered on the Red List of the International Union for the Conservation of Nature because its numbers have declined significantly in recent years, up to 90 percent of the population has been lost in some areas.

Discovering how corals respond to ocean warming is complicated because corals serve as hosts to algae. The algae live in the coral and feed on its nitrogen wastes. Through photosynthesis, the algae then produce the carbohydrates that feed the coral. When this complex and delicate symbiosis is upset by a rise in ocean temperature, the coral may expel the algae in a phenomenon known as coral bleaching, which may cause the death of both algae and coral. The challenge is to figure out why some corals cope with the heat stress better than others.

"We decided to focus on coral larvae because the successful dispersal and settlement of larvae is key to the survival of reefs," explains Baums. "Also, since free-swimming larvae do not yet have symbiotic algae, we can record the expression of different genes in our samples and know that we are looking at the molecular response of the coral itself to heat stress."

Star coral broadcasts eggs and sperm into the water column in mass spawning events, which occur in the Caribbean a few days after the full moon in August. Fertilization occurs quickly when the larvae reach the surface, and then they drift for as much as two weeks before settling on the hard surfaces where they will spend the rest of their lives. Free-swimming larvae are especially vulnerable to ecological changes because they have limited energetic reserves. Scientifically, studying coral larvae has distinct advantages over documenting the response of adult coral to thermal stress.

Logistically, however, studying larvae scientifically is not so easy. "We have to find suitable reefs with known, and therefore roughly predictable, spawning habits," explains Baums. "These reefs have to be close enough to shore that we can get into the water and out to the corals within the first hour of spawning, which always happens at night. When we see that the corals are about to spawn, we set up nets over coral colonies to catch the fragile gametes before they can reach the surface, then we rush back to shore to set up controlled matings and get the young corals back into aquarium tanks before they die." Once spawning started, the scientists worked nearly around the clock for a few days. If they failed to capture enough larvae, or if the larvae died in captivity, the experiment could not be repeated until the following year.

The team successfully collected spawn from two populations of mountainous star coral, one off Key Largo, Florida, and one off Puerto Morales, Mexico. Keeping spawn from the two sites separate, the scientists allowed fertilization to occur in captivity, then they raised the embryos at different temperatures. They recorded the developmental stage and gene expression in the embryos between 12 and 48-50.5 hours after fertilization, comparing those embryos raised at normal temperatures with those raised at temperatures that were 1-2 degrees Centigrade higher.

The embryos from Florida and Mexico developed similarly in the first 50 hours, with the high-temperature embryos maturing only slightly faster than the embryos raised at normal temperatures. Strikingly, larvae raised at higher temperatures showed many more irregular, misshapen embryos than those raised at normal temperatures. For example, after 46 hours, fully 50 percent of the high-temperature embryos from Florida were deformed as compared to the normal-temperature embryos, none of which were malformed. The Mexican samples showed the same pattern but those embryos were less strongly affected by the elevated temperature. Although both populations represent the same species, they responded differently to heat stress, showing genetic variability within the species.

In addition to examining the physical appearance of embryos as they developed, the team extracted RNA from approximately 1,500 embryos from each location to see how much of each of 1,300 molecular products were being transcribed at a given time. Genes that were transcribed in different amounts between high-temperature and normal-temperature samples were called deferentially expressed genes. Twenty-four hours after fertilization, embryos from the same site showed similar gene expression profiles regardless of the temperature at which they were raised.

As the time since fertilization increased, the samples showed more and more deferentially expressed genes, 458 in all. Of the 218 deferentially expressed genes that were sensitive to temperature, almost none were shared between the two locations on the first day of sampling, but by the second day, roughly 25 percent were shared between samples from Mexico and Florida. By 48 hours, thermal stress not sampling location became the dominant factor influencing gene expression. At that point, the gene expression of coral subjected to similar temperatures clustered together regardless of their place of origin.

The team then classified the deferentially expressed genes into functional groups and found that the genes most sensitive to temperature changes were primarily those involved in cell proliferation, growth, and development. The genes that varied according to location of origin were most often involved in cell adhesion, protein degradation, and protein biosynthesis.

"Our study shows that the response of larvae to changing conditions depends upon where the parent colonies lived," says Baums. "Clearly the coral larvae from Mexico and Florida respond differently to heat stress, even though they belong to the same species, showing adaptations to local conditions. The two populations have different adaptive potential."

Baums said she is excited by the clear evidence of local adaptations in populations that this study documented. Previous work by Baums and her colleagues has included experiments in restoring damaged coral reefs by creating larvae from controlled genetic crosses, growing them in captivity until they settle onto ceramic tiles, and then transplanting them into selected areas to replenish damaged reefs. Some crosses survive in higher-temperature water better than others, some survive in captivity better than others, and some settle more reliably onto the prepared tiles that are used to form or restore colonies. The new information from the current study will be invaluable in restoration work.

"Variation among populations in gene expression offers the species as a whole a better chance of survival under changing conditions," Baums said. "We might be able to screen adult populations for their ability to produce heat-resistant larvae and focus our conservation efforts on those reefs

WWF Study Says Climate Change could Displace Millions in Asia's Coral Triangle

Coral reefs could disappear entirely from the Coral Triangle region of the Pacific Ocean by the end of the century, threatening the food supply and livelihoods for about 100 million people, according to a new study from World Wildlife Fund. Averting catastrophe will depend on quick and effective global action on climate change coupled with the implementation of regional solutions to problems of over-fishing and pollution, according to The Coral Triangle and Climate Change: Ecosystems, People and Societies at Risk, a WWF-commissioned study presented at the World Oceans Conference in Manado, Indonesia today.

"This area is the planet's crown jewel of coral diversity and we are watching it disappear before our eyes," said Catherine Plume, Director of the Coral Triangle Program for WWF-US. "But as this study shows, there are opportunities to prevent this tragedy while sustaining the livelihoods of millions who rely on its riches."

The report offers two dramatically different scenarios for the Coral Triangle, which is comprised of the coasts, reefs and seas of the countries of Indonesia, the Philippines, Malaysia, Papua New Guinea, the Solomon Islands and Timor Leste. The Coral Triangle occupies just one percent of the Earth's surface, but is home to fully 30 percent of the world's coral reefs, 76 percent of reef-building coral species and more than 35 percent of coral reef fish species. It is also serves as vital spawning grounds for other economically important fish such as tuna.

"In one scenario, we continue along our current climate trajectory and do little to protect coastal environments from the onslaught of local threats," said Queensland University Professor Ove Hoegh-Guldberg, who led the study. "In this world, people see the biological treasures of the Coral Triangle destroyed over the course of the century by rapid increases in ocean temperature, acidity and sea level, while the resilience of coastal environments also deteriorates under faltering coastal management. Poverty increases, food security plummets, economies suffer and coastal people migrate increasingly to urban areas."

The report also highlighted opportunities to avoid a worst-case scenario in the region through significant reductions in greenhouse gas emissions and international investment in strengthening the region's natural environments, solutions that would help to build a resilient and robust Coral Triangle in which economic growth, food security and natural environments are maintained.

"Climate change in the Coral Triangle is challenging but manageable, and the region would respond well to reductions in local environmental stresses from overfishing, pollution, and declining coastal water quality and health," Hoegh-Guldberg said.

Even under the best case scenario however, communities in the region can expect to experience dramatic losses of coral, rising sea level, increased storm activity, severe droughts and reduced food availability from coastal fisheries. But effective management of coastal resources would mean the communities would remain reasonably intact and more resilient in the face of such hardships.

WWF officials said world leaders have a role to play in helping Coral Triangle countries strengthen management of their marine resources and through international action on climate change.

"We must forge a strong international agreement to bring about sharp reductions in greenhouse gases at the UN Climate Conference at Copenhagen in December," Plume said.

Oceans' Growing Carbon Dioxide Levels May Threaten Coral Reef Fish

Much study has been done on the effects of ocean acidification on coral and shelled animals, but little on how the effects would manifest in other forms of marine life, said Munday, who led the study published Tuesday in the Proceedings of the National Academy of Sciences. "What we wanted to find out was how it affects those that don't have a skeleton on their outside."

The scientists put larval fish in water enriched with various levels of carbon dioxide, whose concentration in the oceans has been rising as a result of rising levels in the Earth's atmosphere. The lowest was 390 parts per million (the current level in the ocean) and the highest 850 ppm (which the scientists estimated would be the carbon dioxide level in the water by the end of the century if current trends continue). Then the scientists allowed each of the larval fish to pick a water source — one that had been scented with a predator's chemical signature or one that was clear of dangerous smells.

They did the experiment twice: once with baby clownfishes raised in captivity and once with wild-caught young damselfishes.

Many coral reef fish can smell nearby predators — a key ability, biologists said, given what an appetising snack larval fish make for rock cod, dottybacks and other larger fish.

"They're kind of like Hershey's Kisses ... everybody's after them," said Mark Hay, a marine ecologist at the Georgia Institute of Technology, who was not involved in the study.

Normally, larval fish would flee from the predator odours. But fish exposed to the highest levels of carbon dioxide in the experiment did not: They even seemed to be attracted to the very odour that should have set off their neuronal alarms.

The scientists then tested these fish in the ocean. They lopped off pieces of coral reef and moved them to empty spots in the sand, making temporary one-fish habitats. Experienced scuba divers discreetly watched each fish as it swam around its new home.

They found that the fish that had spent time in the highest levels of carbon dioxide ventured farther away from their coral and acted much more boldly than their counterparts in normal water — striking aggressively at food and exploring without trying to hide, for example. They were also five to nine times more likely to die.

In other words, merely having been exposed to higher levels of carbon dioxide altered the behaviour of the fish for some time after they were again in water with normal levels.

Hay said the study's findings raised questions — such as how carbon dioxide affects a fish's ability to smell and whether the findings are applicable to many other types of fish and marine life. But he said the study shows that the effects of greenhouse gases on marine life could be far more complex and far-reaching than thought.

Coral Reefs May Start Dissolving when Atmospheric CO_2 Doubles

Rising carbon dioxide in the atmosphere and the resulting effects on ocean water are making it increasingly difficult for coral reefs to grow, say scientists. A study to be published online March 13, 2009 in *Geophysical Research Letters* by researchers at the Carnegie Institution and the Hebrew University of Jerusalem warns that if carbon dioxide reaches double pre-industrial levels, coral reefs can be expected to not just stop growing, but also to begin dissolving all over the world. The impact on reefs is a consequence of both ocean acidification caused by the absorption of carbon dioxide into seawater and rising water temperatures. Previous studies have shown that rising carbon dioxide will slow coral growth, but this is the first study to show that coral reefs can be expected to start dissolving just about everywhere in just a few decades, unless carbon dioxide emissions are cut deeply and soon.

"Globally, each second, we dump over 1000 tons of carbon dioxide into the atmosphere and, each second, about 300 tons of that carbon dioxide is going into the oceans," said co-author Ken Caldeira of the Carnegie Institution's Department of Global Ecology, testifying to the U.S. House of Representatives Subcommittee on Insular Affairs, Oceans and Wildlife of the Committee on Natural Resources on February 25, 2009. "We can say with a high degree of certainty that all of this CO_2 will make the oceans more acidic that is simple chemistry taught to freshman college students."

The study was designed determine the impact of this acidification on coral reefs. The research team, consisting of Jacob Silverman, Caldeira, and Long Cao of the Carnegie Institution as well as Boaz Lazar and Jonathan Erez from The Hebrew University of Jerusalem, used field data from coral reefs to determine the effects of temperature and water chemistry on coral calcification rates. Armed with this information, they plugged the data into a computer model that calculated global seawater temperature and chemistry at different atmospheric levels of CO_2 ranging from the pre-industrial value of 280 ppm (parts per million) to 750 ppm. The current atmospheric concentration is over 380 ppm, and is rapidly rising due to human-caused emissions, primarily through the burning of fossil fuels. Based on the model results for more than 9,000 reef locations, the researchers determined that at the highest concentration studied, 750 ppm, acidification of seawater would reduce calcification rates of three quarters of the world's reefs to less than 20% of pre-industrial rates. Field studies suggest that at such low rates, coral growth would not be able to keep up with dissolution and other natural as well as manmade destructive processes attacking reefs.

Prospects for reefs are even gloomier when the effects of coral bleaching are included in the model. Coral bleaching refers to the loss of symbiotic algae that are essential for healthy growth of coral colonies. Bleaching is already a widespread problem, and high temperatures are among the factors known to promote bleaching. According to their model the researchers calculated that under present conditions 30% of reefs have already undergone bleaching and that at CO_2 levels of 560 ppm (twice pre-industrial levels) the combined effects of acidification and bleaching will reduce the calcification rates of all the world's reefs by 80% or more. This lowered calcification rate will render all reefs vulnerable to dissolution, without even considering other threats to reefs, such as pollution.

"Our fossil-fuelled lifestyle is killing off coral reefs," says Caldeira. "If we don't change our ways soon, in the next few decades we will destroy what took millions of years to create."

"Coral reefs may be the canary in the coal mine," he adds. "Other major pieces of our planet may be similarly threatened because we are using the atmosphere and oceans as dumps for our CO_2 pollution. We can save the reefs if we decide to treat our planet with the care it deserves. We need to power our economy with technologies that do not dump carbon dioxide into the atmosphere or oceans."

Voracious Sponges Save Reef

Tropical oceans are known as the deserts of the sea. And yet this unlikely environment is the very place where the rich and fertile coral reef grows. Dutch researcher Jasper de Goeij investigated how caves in the coral reef ensure the reef's continued existence. Although sponges in these coral caves take up a lot of dissolved organic material, they scarcely grow. However, they do discard a lot of cells that in turn provide food for the organisms on the reef. Caves in coral reefs are the largest and least well known part of the reef. De Goeij investigated coral caves near Curacao and Indonesia. Up until now it had been assumed that cave sponges could only eat by filtering the non-dissolved particles from the seawater. This research demonstrated, however, that the caves contain far more dissolved material than non-dissolved material.

The Reef's Guts

Cave sponges take up enormous quantities of dissolved organic material from seawater. The question is whether they merely take up the material or whether they also process it. De Goeij revealed that the sponges process forty percent of the material and take up sixty percent. This should lead to a doubling of the sponges' biomass every two to three days. However, cave sponges scarcely grow.

The coral caves are densely populated and so there is scarcely any space to grow. Instead of growing the cave sponges rapidly rejuvenate their filtration cells and discard their old cells. This short cell cycle is unique for multicellular organisms and to date was only known to occur in unicellular organisms. The production and breakdown process of the sponge cells mirrors that in the human intestinal tract.

Eating and Being Eaten

Coal reef maintains itself in a remarkable manner. The algae and corals on the reef produce dissolved organic material. Before this material flows into the open ocean sponges in the caves take it up. The sponges rapidly filter enormous quantities of water and convert dissolved material into particles. These particles are in turn consumed by the algae and corals on the reef. In this manner, the various inhabitants of the reef facilitate each other's survival.

Sponges Recycle Carbon to Give Life to Coral Reefs

Coral reefs support some of the most diverse ecosystems on the planet, yet they thrive in a marine desert. So how do reefs sustain their thriving populations? Marine biologist Fleur Van Duyl from the Royal Netherlands Institute for Sea Research is fascinated by the energy budgets that support coral reefs in this impoverished environment. According to van Duyl's former student, Jasper De Goeij, *Halisarca caerulea* sponges grow in the deep dark cavities beneath reefs, and 90% of their diet is composed of dissolved organic carbon, which is inedible for most other reef residents. But when De Goeij measured the amount of carbon that the brightly coloured sponges consumed he found that they consume half of their own weight each day, yet they never grew. What were the sponges doing with the carbon? Were the sponges really consuming that much carbon, or was there a problem with De Goeij's measurements? He had to find out where the carbon was going to back up his measurements and publishes his discovery that sponges have one of the fastest cell division rates ever measured, and instead of growing they discard the cells. Essentially, the sponges recycle carbon that would otherwise be lost to the reef.

Travelling to the Dutch Antilles with his student, Anna De Kluijver, De Goeij started SCUBA diving with the sponges to find out how much carbon they consume. 'It is quite dark and technically difficult to work in the cavities,' explains De Goeij, but the duo collected sponges, placed them in small chambers and exposed the sponges to 5- bromo-22 -deoxyuridine (BrdU). 'The BrdU is only incorporated into the DNA of dividing cells,' explains De Goeij, so cells that carry the BrdU label must be dividing, or have divided, since the molecule was added to the sponge's water, and cells can only divide if they are taking up carbon. But when De Goeij returned to the Netherlands with his samples, he had problems finding the elusive label.

Discussing the BrdU detection problem with his father, biochemist Anton De Goeij, De Goeij Senior offered to introduce his son to Bert Schutte in Maastricht, who had developed a BrdU detection system for use in cancer therapy. Maybe he could help De Goeij Junior find evidence of cell division in his sponges.

Taking his samples to Jack Cleutjens's Maastricht Pathology laboratory, De Goeij was finally able to detect the BrdU label in his sponge cells. Amazingly, half of the sponge's choanocyte (filtration) cells had divided and the choanocyte's cell division cycle was a phenomenally short 5.4 hours. 'That is quicker than most bacteria divide,' exclaims De Goeij.

The sponge was able to take up the colossal amounts of organic carbon that De Goeij had measured, but where was the carbon going: the sponges weren't growing. De Goeij tested to see if the cells were dying and being lost, but he couldn't find any evidence of cell death.

Presenting his results to the Maastricht Pathology Department, someone said 'Lets look at this like a human intestine, then you should see shedding where old cells detach from the epithelia'. De Goeij knew that he had seen some loose cells, and thought that they were artefacts from cutting the samples, but when he and his Pathology Department colleagues went back and looked at the samples, De Goeij realised that choanocytes were shedding all over the place. And then De Goeij remembered the tiny piles of brown material he found next to the sponges in the aquarium every morning.

The sponges were shedding the newly divided cells, which other reef residents could now consume. '*Halisarca caerulea* is the great recycler of energy for the reef by turning over energy that nobody else can use [dissolved organic carbon] into energy that everyone can use [discarded choanocytes],' explains De Goeij

Large Sponges May be Re-attached to Coral Reefs

A new study appearing in *Restoration Ecology* describes a novel technique for re-attaching large sponges that have been dislodged from coral reefs. The findings could be generally applied to the restoration of other large sponge species removed by human activities or storm events.

20 specimens of the Caribbean giant barrel sponge were removed and re-attached at Conch Reef off of Key Largo, Florida in 2004 and 2005 at depths of 15m and 30m. The sponges were affixed to the reef using sponge holders consisting of polyvinyl chloride piping, which was anchored in a concrete block that was set on a plastic mesh base.

Though the test area endured four hurricanes during the study period, 62.5 percent of sponges survived at least 2.3-3 years and 90 percent of the sponges attached in deep water locations survived. The sponges re-attached to the reef after being held stationary by sponge holders for as little as 6 months.

Large sponges may be damaged by a variety of natural events and human activities including severe storms, vessel groundings and the cutting movements of chain or rope moved along with debris by strong currents. After these events, detached large sponges are commonly found, still alive and intact, between reef spurs on sand or rubble where they slowly erode under the action of oscillating currents.

"The worldwide decline of coral reef ecosystems has prompted many local restoration efforts, which typically focus on reattachment of reef-building corals," says Professor Joseph Pawlik of the University of North Carolina-Wilmington, co-author of the study. "Despite their dominance on coral reefs, large sponges are generally excluded from restoration efforts because of a lack of suitable methods for sponge reattachment."

These sponges, which often exceed reef-building corals in abundance, can be more than 1m in diametre and may be hundreds or thousands of years old. The success of past attempts at re-attaching sponges, which used cement or epoxy, has been limited because adhesives do not bind to sponge tissue. When damaged or dislodged, large sponges usually die because they are unable to reattach to the reef. The results of the study show that these sponges have the ability to reattach to the reef if they can be properly secured.

Study Finds Fisheries Management Makes Coral Reefs Grow Faster

An 18 year study of Kenya's coral reefs by the Wildlife Conservation Society and the University of California at Santa Cruz has found that overfished reef systems have more sea urchins, organisms that in turn eat coral algae that build tropical reef systems. By contrast, reef systems closed to fishing have fewer sea urchins, the result of predatory fish keeping urchins under control and higher coral growth rates and more structure.

The paper appears in the December 2010 issue of the scientific journal *Ecology*. The authors include Jennifer O'Leary of the University of California at Santa Cruz and Tim McClanahan of the Wildlife Conservation Society.

The authors found that reefs with large numbers of grazing sea urchins reduced the abundance of crustose coralline algae, a species of algae that produce calcium carbonate. Coralline algae contribute to reef growth, specifically the kind of massive flat reefs that fringe most of the tropical reef systems of the world.

The study focused on two areas—one a fishery closure near the coastal city of Mombasa and another site with fished reefs. The researchers found that sea urchins were the dominant grazer in the fished reefs, where the predators of sea urchins- triggerfish and wrasses were largely absent. The absence of predators caused the sea urchins to proliferate and coralline algae to become rare.

"These under-appreciated coralline algae are known to bind and stabilise reef skeletons and sand as well as enhance the recruitment of small corals by providing a place for their larvae to settle," said Dr. Tim McClanahan, WCS Senior Conservationist and head of the society's coral reef research and conservation program. "This study illustrates the cascading effects of predator loss on a reef system and the importance of maintaining fish populations for coral health."

The study also focused on the effects of herbivorous fish—surgeonfish and parrotfish on coral reefs. While these 'grazing' fish did measurably impact the growth rates of coralline algae in reef systems, they also removed fleshy algae that compete with coralline algae. Overall, reefs with more sea urchins grew significantly slower than ones with more complete fish communities.

The authors also found that the grazing effect was stronger and more persistent than the strong El Niño that devastated coral reefs throughout the tropics in 1998 (the study extended from 1987 until 2005). The study shows that managing coral reef fisheries can affect coral reef growth and improving the management of tropical fisheries can help these reefs to grow and persist in a changing climate.

"The survival of coral reefs is critical for hundreds of millions of people who depend on these complex systems for coastal protection, food, and tourism revenue around the globe," said Dr. Caleb McClennen, Director of WCS's Marine Program. "This study demonstrates the importance of improving fisheries management on reefs so that corals can thrive, safeguarding some of the world's most fragile marine biodiversity and strengthening coastal economies.

Deep-sea Corals May be Oldest Living Marine Organism

Deep-sea corals from about 400 metres off the coast of the Hawaiian Islands are much older than once believed and some may be the oldest living marine organisms known to man. Researchers from Lawrence Livermore, Stanford University and the University of California at Santa Cruz have determined that two groups of Hawaiian deep-sea corals are far older than previously recorded.

Using the Lab's Centre for Accelerator Mass Spectrometry, LLNL researchers Tom Guilderson and Stewart Fallon used radiocarbon dating to determine the ages of Geradia sp., or gold coral, and specimens of the deep-water black coral, Leiopathes sp. The longest lived in both species was 2,740 years and 4,270 years, respectively. At more than 4,000 years old, the deep-water black coral is the oldest living skeletal-accreting marine organism known. "And to the best of our knowledge, the oldest colonial organism yet found," Guilderson said. "Based on the carbon 14, the living polyps are only a few years old, or at least their carbon is, but they have been continuously replaced for centuries to millennia while accreting their underlying skeleton."

The research appears in the March 23 early online edition of the *Proceedings of the National Academy of Sciences.*

Using a manned deep-sea research submersible, the team used samples that were individually collected from the Makapuu and Lanikai deep-sea coral beds off the coast of Oahu, Keahole Point deep-sea coral bed off the coast of the Big Island and Cross Seamount about 100 miles south of Oahu.

Carbon dating uses radiocarbon (carbon 14) to date the age of an object. Radiocarbon is the most widely used geochronological tool in the earth sciences for the late Quaternary (the last 50,000 years). Earlier radiocarbon studies showed that individual gold coral colonies from the Atlantic and Pacific oceans have life spans of 1,800 to 2,740 years, but the results remain contentious with some biologists. In particular, some have questioned whether the corals feed on re-suspended sediment (which could be old) and not on recently photosynthesized carbon that falls through the water column, or that they grew faster and then stopped growing when they reached a certain size.

To answer these questions, the group analysed not only polyps (the living animals that make up corals) but a branch of one specimen.

The living animals had the same carbon 14 concentration as the overlying surface water. This shows that the carbon in the polyps was recently photosynthesized in the surface prior to being "eaten" by the polyps. The skeleton's carbon 14 concentration mimicked that of the overlying surface water's 'post-bomb' time series: the time since the late 1950s when the testing of nuclear weapons augmented the natural abundance of carbon 14 in the atmosphere.

The radial growth rate during the last 50 years is similar to the long-term growth rate of the 300-year branch. The radial growth rate also is consistent with that derived from larger fossil samples. The radial growth rate is similar within a rather small range of tens of microns per year for all specimens analysed.

In the recent research, the Geradia coral was assumed to be much younger when amino acid and growth band methods were used. With radiocarbon dating, the average life span of the analysed specimens is 970 years and ranges from about 300 years for a small branch (with a radius of 11 millimetres) to about 2,700 years (with a radius of 38 mm).

"These ages indicate a longevity that far exceeds previous estimates," Guilderson said. "Many of the Geradia samples that we have analysed are branches, not the largest portions of the colony and so the ages may not indicate how old the entire individual is."

Hawaiian deep sea corals face direct threats from harvesting for jewellery and from commercial fisheries that trawl the ocean bottoms. In addition, the close relationship between deep sea corals (and the mid water ecosystems) and ocean's surface means that they can be affected by natural and manmade changes in surface ocean conditions including ocean acidification, warming and altered stratification.

The antiquity of the coral is an additional call for action, Guilderson said.

"The extremely long life spans reinforce the need for further protection of deep-sea habitat" he said. "The research has already had an impact for activities in Hawaiian waters where a harvesting and fishing moratorium has been enacted to protect certain areas. There are similar habitats in international waters and it is hoped that the results will provide the scientific basis for agreements under the Law of the Sea, and United Nations Environment Programme."

Seamounts May Serve as Refuges for Deep-sea Animals that Struggle to Survive Elsewhere

Over the last two decades, marine biologists have discovered lush forests of deep-sea corals and sponges growing on seamounts (underwater mountains) offshore of the California coast. It has generally been assumed that many of these animals live only on seamounts, and are found nowhere else. However, two new research papers show that most seamount animals can also be found in other deep-sea areas. Seamounts, however, do support particularly large, dense clusters of these animals. These findings may help coastal managers protect seamounts from damage by human activities. Tens of thousands of seamounts dot the world's ocean basins. Although some shallower seamounts have been used as fishing grounds, few seamounts have been studied in detail. Davidson Seamount, about 120 kilometres (75 miles) offshore of the Big Sur coast, is an exception. Since 2000, researchers have spent over 200 hours exploring its slopes and peaks using the remotely operated vehicle (ROV) Tiburon. Two of the expeditions to Davidson Seamount were led by Andrew DeVogelaere of the Monterey Bay National Marine Sanctuary and were funded by the National Oceanic and Atmospheric Administration's Office of Exploration. Other expeditions were funded by the David and Lucile Packard Foundation (through MBARI) and were led by MBARI biologist James Barry, who studies seafloor animals, and by geologist David Clague, who studies undersea volcanoes.

Following each expedition to Davidson Seamount, marine biologists at MBARI studied high-resolution video taken by the ROV and identified every animal they could see. Over 60,000 of these observations were entered in MBARI's video annotation and reference system (VARS). Craig McClain and Lonny Lundsten, the lead authors of the two recent papers, used the VARS database to find out which animals were unique to Davidson Seamount and which had been seen elsewhere. Altogether, 168 different species of animals were observed on Davidson Seamount. McClain's search of the VARS database showed that 88% of these animals had also been seen or reported in other deep seafloor areas, such as the walls of Monterey Canyon. Three quarters of the species on Davidson were not even unique to the California coast, and had been seen in seafloor areas over 1000 kilometres (620 miles) away, including the Hawaiian Islands, the Sea of Japan, and Antarctica.

Only about seven percent of the species at Davidson Seamount had never been seen anywhere else. Of these 12 apparently "endemic" species, most were new to science. Thus, their full ranges are still unknown.

Although few animals are "endemic" to Davidson Seamount, the research demonstrated that this seamount does support distinctive groups of animals, which are dominated by extensive "forests" of large, "old-growth" corals and sponges. These same species of corals and sponges also grow on the walls of Monterey Canyon, but usually as smaller, scattered individuals. Conversely, sea cucumbers are common on the walls of Monterey Canyon, but are rare at Davidson Seamount. Thus, animals that are common on Davidson Seamount are uncommon in other seafloor areas, and vice versa.

The researchers speculate that Davidson Seamount is a good habitat for deep-sea corals and sponges because it has favourable bottom materials (bare lava rock), a steady food supply (drifting particles of the right size and type), and may be less disturbed by strong bottom currents than other seafloor areas. Craig McClain, one of the lead authors, explains, "The large groves of corals and sponges are unique to seamounts. The crests of seamounts are particularly good because they provide flat rocky surfaces that don't accumulate much sediment. This is partly due to the fact that seamounts are so far offshore."

In contrast, McClain points out, "When you look at the seafloor in Monterey Canyon, it's mucky. That makes it tough for filter feeders, especially sponges. Any flat surface in the canyon collects mud. This makes it tough for corals to settle anywhere except on near-vertical surfaces. Just staying attached to these surfaces can be a challenge in itself."

McClain and Lundsten's research also suggests that seamounts such as Davidson Seamount may be ecologically important as breeding grounds for animals that are rare in other habitats. As McClain writes in his paper, "seamounts are likely to be sources of larvae that maintain populations of certain species in sub-optimal, non-seamount sinks." He explains, "Sources are places where certain species do really well—they're self sustaining populations. Sinks are areas where these species can live, but do very poorly. Populations in sink areas will die out if they're not continuously replenished by new animals from source areas." The researchers suggest that future DNA studies of seamount animals would help scientists find out if seamounts are indeed sources of larvae for other seafloor areas.

Lundsten's paper emphasizes the fact that not all seamounts are alike. For example, Rodriguez Seamount, a smaller seamount offshore of Point Conception, once extended above sea level. Thus, Rodriguez Seamount has a flat, sediment-covered crest that is partially covered with ancient beach sands. These sands have been colonised by a very different set of animals from those at Davidson Seamount. In fact, sea cucumbers are the most abundant animals on Rodriguez Seamount.

In 2008, Davidson Seamount was added to the Monterey Bay National Marine Sanctuary. Findings from McClain's and Lundsten's papers will provide critical information for managing Davidson Seamount, and could be useful in other sea-life protection efforts around the world.

Prior to this study, seamounts were considered isolated biological "islands," which might require management to protect certain unique species. This study, on the other hand, suggests that seamounts should be managed as entire communities, whose dense populations of animals release larvae that help colonise other, less optimum environments. Either way, the authors point out, seamounts are well worthy of our protection.

Marine Scientists Unveil the Mystery of Life on Undersea Mountains

They challenge the mountain ranges of the Alps, the Andes and the Himalayas in size yet surprisingly little is known about seamounts, the vast mountains hidden under the world's oceans. Now in a special issue of *Marine Ecology* scientists uncover the mystery of life on these submerged mountain ranges and reveal why these under studied ecosystems are under threat. The bathymetry of our oceans is now resolved at a scale and detail unimaginable by early pioneers and recent estimates suggest that, globally, there may be up to 100,000 seamounts, yet despite best efforts less than 300 have been well studied. Recognising this scarcity of knowledge provided the motivation CenSeam, a seamount focused field within the Census of Marine Life which commenced in 2005.

"The field of seamount ecology is rife with ecological paradigms, many of which have already become cemented in the scientific literature and in the minds of advocates for seamount protection," said Dr Ashley Rowden, one of the principal investigators of CenSeam. "Together, these paradigms have created a widely held view of seamounts as unique environments, hotspots of biodiversity with fragile ecosystems of exceptional ecological worth."

The special issue puts major paradigms in seamount ecology under the microscope to assess their status against the weight of existing evidence to date, and against the backdrop of the latest findings.

Researchers challenged the theory that seamounts act as hotspots of species richness, the weight of evidence now suggests that seamounts may have comparable levels of diversity and endemism to continental margins. However, it appears that their ecological communities are distinct in structure, and of higher biomass than neighbouring continental margins.

The geographical differences between seamount communities have suggested limited larval dispersal, local speciation, geographic isolation, or a combination of all these processes. New genetic research presented in the special issue addresses these themes, documenting complex patterns of connectivity among species populations that depend on spatial scale, physical barriers, and life history characteristics.

Much seamount research has been born out of the need to better manage these potentially vulnerable ecosystems. Globally, seamount ecosystems are under pressure from bottom-contact fishing and other human-related impacts. Researchers detail the footprint of trawling and a risk assessment confirms what has long been suspected: seamount communities are highly vulnerable to disturbance by bottom trawling and recovery from fishing impacts is a lengthy process, likely requiring decades at a minimum.

A predicted shallowing of the aragonite saturation horizon caused by ocean acidification is expected to place deepwater corals at risk, but researchers pose that the summits and upper flanks of seamounts may yet provide a spatial refuge from these impacts.

Seamounts Reach a Pinnacle in Upcoming Issue of Oceanography

Lying beneath the ocean is spectacular terrain ranging from endless chains of mountains and isolated peaks to fiery volcanoes and black smokers exploding with magma and other minerals from below Earth's surface. This mountainous landscape, some of which surpasses Mt. Everest heights and the marine life it supports, is the spotlight of a special edition of the research journal *Oceanography*. These massive underwater mountains, or seamounts, are scattered across every ocean and collectively comprise an area the size of Europe. These deep and dark environments often host a world teeming with bizarre life forms found nowhere else on Earth. More than 99 percent of all seamounts remain unexplored by scientists, yet their inhabitants, such as the long-lived deepwater fish orange roughy, show signs of habitat destruction and over exploitation from intense international fishing efforts.

Scientists from Scripps Institution of Oceanography at UC San Diego and colleagues from the National Oceanic Atmospheric Administration, Oregon State University, University of British Columbia and Woods Hole Oceanographic Institution were among those who contributed their expertise in seamount chemistry, physics, geology, hydrology, oceanography, biology and fisheries conservation to this special interdisciplinary effort to delve into the extremely broad research supported by seamounts and to communicate the science and threats facing them to the public.

"One of the key goals of this special issue was to bring together the extremely diverse seamount research community that ranges from fisheries science and conservation all the way to mantle geochemistry," said Hubert Staudigel a research geologist at Cecil H.

and Ida M. Green Institute of Geophysics and Planetary Physics at Scripps and the lead guest editor of the special issue. "In my eyes, this volume of Oceanography goes beyond that by presenting amazing new research in a way that the public can understand and get excited about."

"This issue of *Oceanography* offers a broad perspective on seamount research of all major disciplines to raise awareness of the diversity of seamount research and to promote collaboration among seamount scientists," wrote the editors of the issue, which represents the most comprehensive volume of peer-reviewed research on the subject to date.

"I was pleased to see how many of the contributions in this special issue deal with very practical and societally important issues of seamounts," said U.S. Geological Survey Director Marcia McNutt.

Glacial Erosion Changes Internal Mountain Structure, Responses to Plate Tectonics

Intense glacial erosion has not only carved the surface of the highest coastal mountain range on earth, the spectacular St. Elias range in Alaska, but has elicited a structural response from deep within the mountain. This interpretation of structural response is based on real-world data now being reported, which supports decades of model simulations of mountain formation and evolution regarding the impact of climate on the distribution of deformation associated with plate tectonics.

A team of researchers from seven universities report the results of their field studies, on the structural response of the St. Elias range to glacial erosion, in *Nature Geosciences**. The paper was a partnership of Aaron L. Berger, whose Ph.D. research it encompassed, his major professor, James A. Spotila, both with the Virginia Tech geosciences department; Sean P.S. Gulick of the Institute for Geophysics, Jackson School of Geosciences, at the University of Texas at Austin; and other colleagues. Berger and Spotila headed the land-based erosion research team. Gulick headed the ocean-based seismic reflection and sedimentation research team. The project is part of the National Science Foundation-funded St. Elias Erosion-Tectonics Project (STEEP), lead by Terry L. Pavlis of the University of Texas, El Paso.

The St. Elias range is a result of 10 million years of the North American plate pushing material up as it overrides the Pacific plate, then the material being worn down by glaciers. A dramatic cooling across the earth about three million years ago resulted in the onset of widespread glaciation. A million years ago, glacial conditions became more intense and glaciers grew larger over longer periods, and transitioned into more erosive ice streams that changed the shape and evolution of the mountains. The process continues today, resulting in the particularly active and dramatic St. Elias "orogen" – geologists' word for mountains that grow from collision of tectonic plates.

"The collisions of tectonic plates over millions of years leave a record in the sediments, but it is a history that is difficult to extract. The signals of the impact of climate are even more difficult to track. Which is why scientists have used mathematical models," said Spotila.

Models create a simplified numeric version of an orogen. Then scientists can change variables in the mathematical formula to determine what happens as a result of climate – whether rain or glaciers. "Models are important in that they showed us that climate change can effect mountain growth," Spotila said. "And the St. Elias orogen behaves very differently than ones that are at lower latitudes and receive most of their precipitation as rain," he said.

Armed with the insight of the models, Spotila, his Virginia Tech students, and colleagues at other universities have braved the mountain over many years to collect physical evidence. They have been dropped in remote and dangerous locations by helicopter to place instruments and collect samples to determine bedrock cooling rates and sedimentation. "But our data set wouldn't have shown the complete picture," said Spotila. "We looked at the erosion history onshore and Gulick's team looked at the record off shore – the shelf where the eroded sediment rest." Offshore seismic and borehole data indicate that the increase in offshore sedimentation corresponds to a one-million-year ago change in glaciation and deformation.

How does a change of the mountain surface result in a change of its internal structure? Spotila explained, "If you push snow with a plow, it will always pile up in front of the plow with the same shape," called the Coulomb wedge when applied to the making of mountains. As the North America plate slips over the Pacific plate, it piles up material for the St Elias orogen with a short side toward the plow inland and a long slope down to the ocean, with the toe dipping into the sea. Enter the glacier. As glacial conditions took hold across the St. Elias orogen, the landscape began to be defined by glacial landforms left on its surface. However, the more extreme glacial cycles, and associated increased erosion, of the last million years pushed the orogen to a tipping point, beyond which the orogen was forced to totally restructure itself, Berger said. There are deformation zones where as much as half of the wedge was removed, the researchers report in the journal article.

Due the onset of accelerated glacial erosion, the St. Elias orogen struggled to maintain its wedge shape. "Rock faulting and folding has become more intense as the orogen internally deforms to adjust to the intensified erosion," said Spotila. "The flux of rock from the mountains to the sea is increasing dramatically."

Berger uses an analogy of a bulldozer pushing sand across the ground. "As the glaciers erode the top of the mountains (the top of the pile of sand), the orogen or entire body of sand, begins readjusting itself internally to maintain its wedge-shape. If you could remove the glaciers and watch the process, the flank of the mountain range where the largest glaciers are located would begin to get planed away by erosion, reducing mean elevation. The removal of this rock would change the local tectonic stress fields, resulting in focused deformation, which would begin to push the mountains back up to replace the eroded material.".

The research showed how a change in climate led to a change in the way the motion of tectonic plates is accommodated by structural deformation within the orogen, Spotila said. "The wedge is still present but has narrowed with the eroded material deposited across the toe. Some faults, which previously responded to the push of the plow or tectonic plate, are relocated to respond to the erosion." Spotila concludes, "It is remarkable that climate and weather and the atmosphere can have such a profound impact on tectonics and the behaviour of the solid earth."

3

Marine Ecology

Marine Biology

Study of life along the seashore, which became known as marine biology by the twentieth century, was first developed and institutionalised in the United States at the end of the nineteenth century. Two distinct traditions contributed to its modern disciplinary form.

Figure: *Marine Biology*

First to emerge was marine biology as a summertime educational activity, chiefly designed to instruct teachers of natural history about how to study nature within a natural setting. The notion was first suggested to Louis Agassiz, the Harvard zoologist and geologist, by his student Nathaniel Southgate Shaler. Shaler had conducted highly successful summer field experiences for geology students, and felt that similar experiences could be valuable for biology students. Encouraged by his wife, Elizabeth—a longtime advocate for educational opportunities for the largely female teaching community—Agassiz obtained funding and opened the Anderson School of Natural History in 1873 on Penikese Island, located not too distant from Cape Cod. Following this school, several others offered similar experiences. The Summer School of the Peabody Academy of Sciences (Salem, Massachusetts) sponsored instruction for teachers in marine botany and zoology in 1876, and the Boston Society of natural History, with the support of the Women's Education Association (WEA) of Boston, started its summer station north of Boston at Alpheus Hyatt's vacation home in Annisquam.

The second tradition was European, where several marine stations operated by 1880, most notably the Stazione Zoologica in Naples. This marine biology laboratory was founded by Anton Dohrn in 1872. The "Mecca for marine biology," as Naples was soon known, attracted scholars from throughout the world. Agassiz's son, Alexander Agassiz, imported Dohrn's notion to his summer home near Newport, Rhode Island, offering the latest microscopical tools for researchers. William Keith Brooks, a student of the elder Agassiz, accepted the invitation and completed his doctoral research with the younger Agassiz in 1875. Then, when Brooks obtained a position at America's first graduate university, Johns Hopkins University, one of his first tasks was to create a research laboratory in marine biology. Thus, the Chesapeake Zoological Laboratory was opened in 1878

The first U.S. marine biology laboratory to incorporate both traditions was the Marine Biological Laboratory (MBL), which opened in Woods Hole, Massachusetts, in 1888. It originally offered courses in marine botany and marine zoology for beginning students and teachers. But its original director, C. O. Whitman, had spent time at Naples and, like his colleague Brooks, wanted to create research opportunities in marine biology for more advanced students and researchers. To accomplish the task, Whitman initiated advanced courses in embryology, invertebrate zoology, cytology, and microscopy, all of which began to attract more sophisticated students. By the early twentieth century, the MBL welcomed only advanced students and investigators.

Similar marine biology laboratories were founded on the Pacific Coast. Stanford University established the Hopkins Marine Station in Pacific Grove, California, in 1892. To the north, the University of Washingt on opened a marine station near Friday Harbour (San Juan Islands, Washington) in 1904. Henry Chandler Cowles, an ecologist from the University of Chicago who had done pioneering studies on the sand dunes of Lake Michigan started a course in intertidal ecology, the first such course in the United States.

One additional West Coast laboratory played a critical role in defining the new field of marine biology, albeit by exclusion. William Emerson Ritter, an embryologist from Berkeley, created a laboratory near San Diego, initially named the San Diego Marine Biological Laboratory, in 1903. But Ritter was interested in a more global approach to investigations by the seashore, an approach he never successfully defined. He was successful, however, in attracting the financial resources of the Scripps family, and soon the Scripps Institution for Biological Research was built north of the village of La Jolla. Ritter specifically stated that he had no intention of forming another MBL on the West Coast, preferring to emphasize a comprehensive study of the sea. After he retired, without creating an educational base for the institution similar to the other stations, he was replaced by Thomas Wayland Vaughan in 1924. The La Jolla station was renamed the Scripps Institution of Oceanography, and marine biology disappeared as a focus.

The three major American marine biology stations throughout the twentieth century and into the twenty-first century are the MBL, Hopkins Marine Station, and Friday Harbour Laboratories. By the end of world War I (1914–1918), the stations defined marine biology as the study of life in the littoral zone (also known as the inter-tidal zone), or the area that serves as an interface between the marine and terrestrial environments. Courses at the laboratories helped to divide marine biology into several speciality areas, including invertebrate zoology, ecology, algology, embryology, and invertebrate physiology. Following World War II (1939–1945), this focus shifted somewhat as more research funding was available in the biological sciences, especially in terms of research questions with an application to medicine and to the exciting field of molecular biology.

Woods Hole's MBL, for example, has all but abandoned the traditional areas of marine biology for specialised medical and genetic research. Most investigations at the MBL by the end of the twentieth century were laboratory-based studies of cellular and molecular processes, with little fieldwork or studies of marine life. At the same time, largely because the West Coast has a more robust intertidal fauna and flora that is largely unaffected by human intervention, Hopkins and Friday Harbour retain a traditional focus on marine biology.

For the most part, marine biology does not include investigations of the open seas, studies of freshwater marine systems, or inquiries into the country's fisheries. Biological oceanography, a subdiscipline of Oceanography, examines biological questions in the oceans, including studies of marine mammals, marine fisheries, and freshwater sources for the ocean (limnology).

Types of Marine Bacteria

Alice Bellone has been a writer since 2005. She has published poetry for "Hull Newsweekly" and was feature editor for her school newspaper, "The Spinnaker." She holds a Bachelor of Arts in English from Hobart and William Smith Colleges in Geneva, N.Y., and studied creative writing at Trinity University in Carmarthen, Wales, U.K.

Marine bacteria are very diverse estimates range from 1,000 to 10,000 types per millilitre of water.

Marine microbiology is the study of microorganisms living in the sea. Among marine microorganisms are bacteria, protozoa, algae, yeasts and molds. Marine bacteria are an important part of marine microbiology because they contribute to the sea's life and energy cycles.

Heterotrophic Bacteria

Heterotrophic bacteria are marine microorganisms responsible for heterotrophic decomposition and re-mineralisation of organic matter. According to Biology, a heterotroph is a life-form that uses organic compounds as a source of carbon. These bacteria dissolve organic carbons by microbial growth, turning it into particulate microbial matter. This process is known as the microbial loop. A variety of organic material is decomposed by marine heterotrophic bacteria, including oil and related hydrocarbons.

Photoautotrophic Bacteria

These marine organisms are capable of synthesizing their own food from inorganic substances, using light as an energy source. They can survive only in environments where both hydrogen sulfide and light occur.

Chemoautotrophic Bacteria

These marine organisms obtain nourishment through oxidation of inorganic chemical compounds, and turn them into organic compounds. They are dependent upon the availability of oxygen and a suitable inorganic electron source.

Biology and Microbiology

These are exciting times in biology, and San Diego State University is a stimulating place for biological study. Recent breakthroughs in genetic engineering, genomics, immunology, and new ideas in evolution, systematics and ecology have all contributed to an atmosphere of enthusiasm and innovation. These subjects and many more can be found in the research programs in the Biology Department, where the curriculum spans the entire spectrum of life, from DNA to entire evolutionary lineages, and from viruses to fir trees to whales.

Microbiology is the study of bacteria, viruses, yeasts, molds, algae and protozoa. These microorganisms are important in their effect on the health and well being of all living creatures, including humans, and they are the key to the biological revolution in genetic engineering. In addition, they are vital to environmental science, food production, and the marine environment.

Ocean Microbes Combat Plastic Pollution

Researchers are investigating microbes that break down plastics as a means of controlling ocean pollution.

Figure: *Bags in Ocean Plastic on Coral.*

Marine microbes that cling to plastics could offer a solution to ocean pollution, according to Jesse Harrison, presenting his research at the Society for General Microbiology's spring meeting in Edinburgh today. The researchers from the University of Sheffield and the Centre for Environment, Fisheries and Aquaculture Science have shown that the combination of marine microbes that can grow on plastic waste varies significantly from microbial groups that colonise surfaces in the wider environment.

This raises the possibility that the plastic-associated marine microbes have different activities that could contribute to the breakdown of these plastics or the toxic chemicals associated with them. Plastic waste is a long-term problem as its breakdown in the environment may require thousands of years. "Plastics form a daily part of our lives and are treated as disposable by consumers. As such plastics comprise the most abundant and rapidly growing component of man-made litter entering the oceans," said Jesse Harrison.

Over time the size of plastic fragments in the oceans decreases as a result of exposure to natural forces. Tiny fragments of 5 mm or less are called 'microplastics' and are particularly dangerous as they can absorb toxic chemicals which are transported to marine animals when ingested.

While microbes are the most numerous organisms in the marine environment, this is the first DNA-based study to investigate how they interact with plastic fragments.

The new study investigated the attachment of microbes to fragments of polyethylene – a plastic commonly used for shopping bags.

The scientists found that the plastic was rapidly colonised by multiple species of bacteria that congregated together to form a 'biofilm' on its surface. Interestingly, the biofilm was only formed by certain types of marine bacteria.

The group, led by Dr Mark Osborn at Sheffield, plans to investigate how the microbial interaction with microplastics varies across different habitats within the coastal seabed research which they believe could have huge environmental benefits.

"Microbes play a key role in the sustaining of all marine life and are the most likely of all organisms to break down toxic chemicals, or even the plastics themselves," suggested Mr Harrison.

"This kind of research is also helping us unravel the global environmental impacts of

Scripps Oceanography Research Pegs ID of Red Tide Killer

Researchers at Scripps Institution of Oceanography at UC San Diego have identified a potential "red tide killer." Red tides and related phenomena in which microscopic algae accumulate rapidly in dense concentrations have been on the rise in recent years, causing hundreds of millions of dollars in worldwide losses to fisheries and beach tourism activities. Despite their wide-ranging impacts, such phenomena, more broadly referred to as "harmful algal blooms," remain unpredictable in not only where they appear, but how long they persist. New research at Scripps has identified a little-understood but common marine microbe as a red tide killer, and implicates the microbe in the termination of a red tide in Southern California waters in the summer of 2005.

While not all algal outbreaks are harmful, some blooms carry toxins that have been known to threaten marine ecosystems and even kill marine mammals, fish and birds.

Using a series of new approaches, Scripps Oceanography's Xavier Mayali investigated the inner workings of a bloom of dinoflagellates, single-celled plankton, known by the species name Lingulodinium polyedrum. The techniques revealed that so-called Roseobacter-Clade Affiliated ("RCA cluster") bacteria several at a time attacked individual dinoflagellate by attaching directly to the plankton's cells, slowing their swimming speed and eventually killing them.

Using DNA evidence, Mayali matched the identity of the RCA bacterium in records of algal blooms around the world.

In fact, it turns out that RCA bacteria are present in temperate and polar waters worldwide. Mayali's novel way of cultivating these organisms has now opened up a new world of inquiry to understand the ecological roles of these organisms. The first outcome of this achievement is the recognition of the bacterium's potential in killing red-tide organisms.

"It's possible that bacteria of this type play an important role in terminating algal blooms and regulating algal bloom dynamics in temperate marine waters all over the world," said Mayali.

The research study, which was coauthored by Scripps Professors Peter Franks and Farooq Azam, is published in the May 1 edition of the journal Applied and Environmental Microbiology.

"Our understanding of harmful algal blooms and red tides has been fairly primitive. For the most part we don't know how they start, for example," said Franks, a professor of biological oceanography in the Integrative Oceanography Division at Scripps. "From a practical point of view, if these RCA bacteria really do kill dinoflagellates and potentially other harmful algae that form dense blooms, down the road there may be a possibility of using them to mitigate their harmful effects."

The researchers based their results on experiments conducted with samples of a red tide collected off the Scripps Pier in 2005. Because RCA bacteria will not grow under traditional laboratory methods, Mayali developed his own techniques for identifying and tracking RCA through highly delicate "micromanipulation" processes involving washing and testing individual cells of Lingulodinium. He used molecular fluorescent tags to follow the bacteria's numbers, eventually matching its DNA signature and sealing its identity.

"The work in the laboratory showed that the bacterium has to attach directly to the dinoflagellate to kill it," said Mayali, "and we found similar dynamics in the natural bloom."

Franks said he found it a bizarre concept of scale that Lingulodinium dinoflagellates, which at 25 to 30 microns in diametre are known to swim through the ocean with long flagella, or appendages, are attacked by bacteria that are about one micron in size and can't swim.

"It's somewhat shocking to think of something like three chipmunks attaching themselves to an elephant and taking it down," said Franks.

While the RCA cluster's role in the marine ecosystem is not known, Azam, a distinguished professor of marine microbiology in the Marine Biology Research Division at Scripps, said harmful algal blooms are an important problem and consideration must be given to the fact that red tide dinoflagellates don't exist in isolation of other parts of the marine food web. Bacteria and other parts of the "microbial loop" feed on the organic matter released by the dinoflagellates and in turn the dinoflagellates are known to feed on other cells (including bacteria) when their nutrients run out.

Dinoflagellate interactions with highly abundant and genetically diverse bacteria in the sea have the potential to both enhance and suppress bloom intensity—but this important subject is only beginning to be explored.

"The newly identified role of RCA cluster is a good illustration of the need to understand the multifarious mechanisms by which microbes influence the functioning of the marine ecosystems," Azam said. "This type of discovery is helping us understand algal bloom dynamics and the interactions among the components of planktonic ecosystems in ways that we'd imagined but previously lacked evidence," said Franks.

Sea Slugs Generating Green Energy

Photosynthesizing sea slugs take 'you are what you eat' to an extreme: by eating photosynthesizing algae, these "solar-powered" sea slugs are able to live off photosynthesis for months. How does this work? Is this just a straightforward case of symbiosis between algae and sea slugs?

It turns out that this is not a case of symbiosis: this is a case of the amazing and ubiquitous power of viruses to dramatically reshape the genetic landscape.

Cheap, powerful sequencing has enabled scientists to sample the viral world like never before. Our understanding of marine viruses, in particular, has exploded as researchers have sequenced whatever they can find in samples of seawater.

These sequencing studies have discovered "cyanophages" - viruses that infect cyanobacteria, bacteria which do much of the world's photosynthesis. These cyanophages very commonly carry photosynthesis genes, which means that opportunities for abound for a marine organism, which is normally incapable of photosynthesis, to pick the genes necessary to maintain chloroplasts, the cellular entities that enable plants, algae, and ~~cyanobacteria~~ to generate energy from sunlight.

There is, however, a high barrier to overcome before an organism can pick up genes from a bacterial virus. Typically, bacterial viruses don't infect eukaryotes, and eukaryotic viruses don't infect bacteria, due to the fact that viruses have to be niche players: they need to specialise, when it comes to manipulating cellular machinery for their own reproduction. They focus in on the unique features of their targets - in a way, they parallel computer viruses, which in most cases are OS-specific.

The most amazing example of virus-mediated gene transfer is the case of the sea slugs that produce green energy: sea slugs which eat photosynthesizing algae, and then live off photosynthesis for months. As far as I know, these are the only animals that are capable of photosynthesis, and, incredibly, there are multiple sea slug species that do this. What happens is this: sea slugs eat algae, and, in a process named *kleptoplasty*, take up the algal chloroplasts by absorption though the gut. The sea slugs then use those stolen chloroplasts to carry out photosynthesis. However, there's a catch: you can't maintain chloroplasts unless your own genome encodes chloroplast genes. Chloroplasts, being descendants of ancient bacterial symbionts, do have their own genomes, but their genomes do not encode all of the genes necessary for chloroplast maintenance: some contribution from the host is required. In the case of plants, algae, key chloroplast genes are found in the nuclear genome of the host.

So what about the sea slug? Animals do not carry any genes for photosynthesis. Or do they? It turns out that photosynthesizing sea slugs have managed to pick up some chloroplast genes (as discovered by sequencing portions of sea slug genomes), and these chloroplast genes show, in their surrounding DNA sequence, the hallmarks of viral transfer.

Furthermore, viral particles have been detected in both the algal chloroplasts and the cells of the sea slugs, which means that these sea slugs are constantly exposed to an opportunity for gene transfer. A few lucky gene swaps occurring over the course of a few million years of chloroplast virus exposure was apparently enough to produce sea slugs capable of photosynthesis. It's not clear what kind of viruses these are, but it's not unreasonable to suppose that they are a particular niche virus species that has gained the capability to infect both algal chloroplasts and sea slug cells.

Photosynthesizing sea slugs are just one example of the amazing gene-shuffling power of viruses. Some of the most important tumour-promoting genes (at least scientifically important, if not always clinically important), like Src, Ras, and Myc, were first discovered in viruses. These genes are not viral genes; they originated in animals, but were subsequently picked up by viruses. Since viruses shuffle in and out of genomes so frequently, it's no surprise that, over millions of years, they've had a big impact on the world's genetic landscape.

A Case for Biohydrogen

The hydrocarbon economy is faltering as oil reserves dwindle worldwide (Hirsch, 2008). Commodity prices have begun to fluctuate drastically due to the uncertain cost of petroleum, which resulted in food riots around the world in 2008. With a steadily decreasing energy supply and the demands on energy systems continually growing, the planet is in dire economic, geopolitical, and environmental straits. In order to halt the advance of climate change, prevent ecological collapse, rescue the global economy, and ensure our energy security, humanity must find a way to harness currently available (non-fossilised) energy. The largest source of energy on Earth, excluding the future potential for thermonuclear fusion reactors, is the sun. Human civilization consumes 15 TW annually while approximately 80,000 TW of solar energy fall on the Earth's surface each year (Makarieva et al., 2008). For hundreds of millions of years, this solar flux has been the driving force for life on Earth.

It is estimated that photosynthesis absorbs and distributes seven times more energy to the biosphere (~100 TW) than is consumed anthropogenically each year (Makarieva et al., 2008). There are many promising technologies in development that will cheaply harness solar output, like next generation photovoltaics, advanced wind turbines, tidal power systems, wave energy generators, solar thermal collectors, biofuels, solar fuels, electrofuels, etc. In the end, some combination of these solutions, coupled with improved energy and transportation infrastructure, will be needed to resolve our energy dilemma. In this article, I focus on cyanobacteria and their potential for hydrogen production, which, if properly developed, could provide sustainable fuel for transportation and industry in a post-petroleum world.

Cyanobacteria and the Rise of Oxygenic Photosynthesis

Approximately four billion years ago, after the formation of Earth, the first signs of life began to appear, scratching out a lithotrophic existence in an oxygen-starved atmosphere (Towe, 1996; Cleaves et al., 2008). Anoxygenic photosynthesis probably developed very early in evolution (Pierson and Olson, 1989), harnessing the power of the sun, but the real metabolic revolution didn't occur for another billion or so years, when photosynthetic eubacteria evolved the ability to split water into electrons, protons and gaseous oxygen (Nisbet et al., 2007). This new metabolic mode, termed oxygenic photosynthesis, effectively gave living organisms access to an endless supply of electrons from water and became such a prolific process that it filled Earth's atmosphere with oxygen, fundamentally transforming the biogeochemistry of the planet (Wille et al., 2007). Moving electrons from chemically stable water molecules to high-energy carbon-carbon bonds spanned an immense reduction-oxidation (redox) gap, powered by a reliable stellar wind of solar electromagnetic radiation, which ultimately increased the size and scope of life and made the development of structurally complex eukaryotic (multicellular) organisms possible (Blankenship and Hartman, 1998). Almost all ecosystems on Earth today are directly or indirectly dependent on oxygenic photosynthesis.

The subjects of this Earth-transforming story, and the only organisms known to have evolved oxygenic photosynthesis, are the cyanobacteria (Nisbet et al., 2007). For a billion years, the electrons in water sat untouched by photosynthetic life because it was energetically unfeasible to oxidise water and reduce $NADP^+$ (an electron carrier needed for the reduction of inorganic carbon) in a single step. In an evolutionary leap, cyanobacteria pioneered the coupling of two photosynthetic reaction centres, P680 and P700 (named for the wavelengths of light they optimally absorb), referred to as photosystem II (PSII) and photosystem I (PSI), respectively (Allen and Martin, 2007; Mimuro et al., 2008). Thus, the redox problem was resolved by first pumping low-energy electrons up the redox gradient to a temporary reservoir (the quinone pool/electron transfer chain), from which the second photosystem (PSI) could pull out excited electrons in order to transfer them to an even higher energy state, sufficient for the reduction of $NADP^+$ to NADPH + H^+.

The rising concentration of oxygen in the atmosphere presented new challenges to early life on Earth. Most species at the time were obligate anaerobes or microaerophiles, and only those organisms that had the ability to deal with highly reactive oxygen radicals and persist in an oxygen-enriched atmosphere would inherit the majority of Earth's habitats (Brioukhanov and Netrusov, 2007; Bendall et al., 2008). Cyanobacteria, in addition to overcoming the general toxicity of oxygen (a byproduct of their new metabolism), had to find a way to acquire nitrogen under aerobic conditions (fixing atmospheric nitrogen gas into biologically-available ammonia).

The molybdenum-iron nitrogenase enzyme is highly conserved among diazotrophs (nitrogen-fixers) and is very sensitive to oxygen, which destroys the activity of the enzyme. Cyanobacteria got around this issue by separating oxygenic photosynthesis and nitrogen fixation, either spatially or temporally (Tsygankov, 2007). Some cyanobacteria only activate nitrogen fixation under dark anaerobic conditions, when PSII is unable to evolve oxygen.

Other cyanobacteria, like filamentous *Nostoc punctiforme*, form a specialised cell type for nitrogen fixation, called a heterocyst. Heterocysts deactivate their PSII complexes, grow thickened cell walls, and exhibit higher intracellular respiration rates, which keep oxygen levels very low (Cardona, 2009). Vegetative cells provide the heterocysts with carbohydrates, while the heterocysts provide the vegetative cells with fixed nitrogen (Cardona, 2009).

Hydrogen Metabolism in Cyanobacteria

Nitrogenase is responsible for the following reaction: N_2 + 8 H^+ + 8 e" + 16 ATP '! 2 NH_3 + H_2 + 16 ADP + 16 P_i (Tikhonovich and Provorov, 2007). This process is very energy-intensive, requiring 8 moles of ATP for every mole of ammonia produced. In addition to ammonia, the nitrogenase enzyme also produces a molecule of hydrogen gas for every molecule of gaseous nitrogen that it fixes. Hydrogen production via nitrogenase is relatively inefficient, due to the large amounts of ATP required, but it can still be used to produce measurable amounts molecular hydrogen.

One impediment to producing hydrogen in this way is the presence of uptake hydrogenases (Schütz et al., 2004), which oxidise nitrogenase-produced hydrogen in order to minimise the energy loss from N-fixation by reclaiming ATP via the oxyhydrogen reaction, removing oxygen from the interior of the cell and providing reducing equivalents for other cellular processes (Tamagnini et al., 2007). Concordantly, uptake-hydrogenase enzyme activity has been shown to correlate with nitrogenase activity (Schütz et al., 2004). Therefore, it is not surprising that a *Nostoc* uptake-hydrogenase knockout mutant exhibits a significantly higher hydrogen output than the wild-type (Lindberg et al., 2004).

Bidirectional hydrogenases, as their name implies, are capable of reversible catalysis of hydrogen formation or consumption ($2H^+ + 2e^- <=> H_2$). Unlike the uptake-hydrogenases, the activity of these enzymes is independent of nitrogenase activity (Schütz et al., 2004). These enzymes seem to act as putative escape valves for the excess reducing power that could build up during metabolism, but their exact function is still under debate (Tamagnini et al., 2007). Cyanobacterial bidirectional hydrogenases have a nickel-iron active site. These nickel-iron hydrogenases are more aerotolerant, but less productive than the iron hydrogenases found in other anaerobic eubacteria and in eukaryotic green algae (Ghirardi et al., 2007). Both nickel-iron and iron hydrogenases require maturation proteins in order to attain catalytic activity, but it has also been suggested that certain [Fe-Fe] hydrogenases do not require a maturation process (Asada et al., 2000; King et al., 2006). These maturation proteins are involved with the insertion of metal clusters into the active site of the hydrogenases (Fontecilla-Camps et al., 2009). Hydrogenases are much more efficient hydrogen producers than nitrogenases, and hold great promise for future biotechnological applications (Tamagnini et al., 2007).

The overall physiology of heterocystous cyanobacteria is highly complex. Advances in genomics, transcriptomics, proteomics and metabolomics are helping scientists develop a holistic view of hydrogen metabolism (Cardona, 2009). In terms of the obstacles facing photobiological hydrogen production, they can be broken down into a few basic issues (Tamagnini et al., 2007). The first main hurdle is that the most prolific hydrogen-producing enzymes, the [Fe-Fe] hydrogenases, are highly oxygen sensitive. For a photobioreactor to be cost-effective, it should produce hydrogen under atmospheric conditions. Some investigations have looked into finding or engineering oxygen-tolerant hydrogenases (Ghirardi et al., 2007). The difficulty with oxygen-tolerant hydrogenases is that their hydrogenase activity has so far been inversely proportional to their level of oxygen tolerance ([Ni-Fe] hydrogenases).

In heterocystous cyanobacteria, this obstacle has been overcome by creating an anoxic microenvironment inside the heterocyst. Another complication arises due to the competition of different metabolic pathways for electrons from PSI. Most of the reducing power generated via oxygenic photosynthesis is diverted towards carbon and nitrogen fixation in *Nostoc punctiforme*. The goal is to better insulate energy-yielding pathways from competing metabolic processes (Agapakis et al., 2010).

Dr. Matthias Rögner, from Ruhr-Universität in Bochum, estimated at a recent conference in Sigtuna, Sweden, that under optimal growth conditions ~75% of all electrons coming from PSI could potentially be diverted to hydrogen production without negatively impacting the organism. While metabolic engineering can be vastly complex, scientists can theoretically simplify this problem by catching the electrons at their source (PSI). PSI hands its electrons off to a ferredoxin, which then shuttles the electrons away to various pathways (Tamagnini et al., 2007). One idea is to physically link a ferredoxin ligand to a hydrogenase, in order to compete with the native ferredoxins and snatch electrons away from alternate pathways and transfer them directly to the hydrogenase (Agapakis et al., 2010).

The Ideal Hydrogen-producing Organism

Nostoc punctiforme was isolated from a mutualistic association with a cycad (Costa and Lindblad, 2002). *Nostoc*, when living symbiotically, has a higher heterocyst frequency than it does as a free-living organism, and will devote much of its metabolism to the production and secretion of specific nitrogen-rich metabolites that are beneficial to its plant host (Enderlin and Meeks, 1983). Since *Nostoc* has evolved to mass-produce a specific metabolite when living symbiotically, it seems almost pre-programmed for fuel production. Heterocystous N-fixing cyanobacteria have minimal nutritional requirements, high photosynthetic efficiencies and can create anoxic microenvironments inside specialised cells that allow anaerobic processes to occur under aerobic conditions. All of these attributes are conducive to affordable bioreactor design for hydrogen production. If *Nostoc* hydrogen metabolism can be effectively engineered, using novel synthetic biology tools, the door to developing successful photobiological hydrogen-yielding technologies is opened.

Synthetic Biology: An Emerging Field

An astounding result of recent genomic sequencing projects is that the length of a genome does not predict the morphological or physiological complexity of an organism. For example, length of the human genome is similar to that of the fruit fly (Mukherji and van Oudenaarden, 2009). Instead, it has been found that biological modularity can help explain the diversity of form and function in the natural world (Mukherji and van Oudenaarden, 2009). A limited subset of predictable biological "parts" can be assembled in various ways to produce molecular "devices", which can be arranged into "systems" to carry out different functions. In this light, one can view a living cell as a combination of co-regulated genetic circuits, working in tandem. Another surprising discovery made recently shows the inherent resiliency of biological circuitry to rewiring.

Isalan et al. (2008), in order to test the limits of perturbing regulatory networks in biological systems, found that randomly rewiring *Escherichia coli* transcriptional networks by synthetically altering transcription factor/promoter pairings only resulted in faulty growth in 5% of cases. This level of tolerance to random restructuring of biological networks is highly conducive to the adaptation and evolvability of living systems by allowing large-scale alterations to be made to an organisms genome without significantly impeding its growth (Isalan et al., 2008; Mukherji and van Oudenaarden, 2009). Biological modularity and network resiliency provide powerful mechanisms for the rapid evolution of novel or optimised processes by reshuffling pre-existing genes/proteins (Isalan et al., 2008; Mukherji and van Oudenaarden, 2009; Picataggio, 2009).

As our understanding of biology grows in breadth and in depth, we come closer to the goal of being able to rationally design biological systems. Thanks to the enormous accumulation of whole-system biological data and the discovery of the modular nature of genetic and enzymatic elements during the past few decades, coupled with advances in *in silico* data analysis/modelling and rapid *in vitro* DNA synthesis technology, a new field, called synthetic biology, has emerged (Picataggio, 2009).

Synthetic biology allows for the rational manipulation of microbial phenotypes by combining systems biology, bioinformatics, protein design and engineering. With a comprehensive understanding of the molecular regulation of gene expression and protein function, biologists can begin to assemble a toolbox of reliable promoters, repressors, activators, ribosomal binding sites, reporters, signalling devices and enzymes, that can be used to design metabolic circuits in a cellular chassis - an autonomous self-replicating framework, or superstructure, that acts as the platform for synthetic circuitry (Picataggio, 2009). Standardisation of genetic tools will streamline process engineering and expand the potential to quickly develop microbial systems for the production of renewable fuels and high-value molecules (Picataggio, 2009).

Synthetic biologists have already succeeded in characterising thousands of biological parts with defined functions and performance parameters, which can be accessed by the Massachusetts Institute of Technology, but they are not yet capable of engineering whole biological systems with the same precision and reliability that, say, electrical engineers are accustomed to. One of the main challenges is to identify the subset of genes that are absolutely necessary for the survival of a minimal genome - the smallest number of genes that allows for the replication of an organism in a particular environment (Cho et al., 1999). Until researchers can build a living cell from the ground up, they won't totally understand the limits of metabolic engineering. The immense potential for engineering crucial synthetic metabolic circuits has already been demonstrated by the work of Dr. Jay Keasling, who produced an important precursor to the antimalarial drug artemisinin in *E. coli* (Hale et al., 2007). Keasling's work has cut the cost of artemisinin by ten fold and will provide many people in the third world with access to crucial malaria treatments, literally saving *millions* of lives

E. coli is an ideal cellular chassis for molecular reprocessing of low-value substrates into high-value products, but in order to use synthetic biology to tackle humanity's energy needs, it is important that we move away from heterotrophic organisms and focus on developing a photosynthetic chassis that can harness the power of the sun. Approaching solar fuel production via direct synthesis from sunlight avoids the drawbacks of traditional fermentation-based methods whose biomass feedstocks often compete directly with food crops (Tenenbaum, 2008). Researchers at UC Davis have recently developed efficient synthetically-derived fuel (isobutyraldehyde, which can be converted to isobutanol) from cyanobacteria (Atsumi et al., 2009). This result is encouraging, and suggests that the field of photosynthetic fuels will continue to grow. In the absence of a minimal genome, heterocystous cyanobacteria seem to be ideal chassis for hydrogen production.

The Development of Synthetic Biology Tools for Cyanobacteria

The creation of standard biological parts (promoters, ribosomal binding sites, repressors, activators, etc.) for cyanobacteria is simplified by the work already completed in *E. coli* and by the current open-source nature of synthetic biology resources, fostered by organisations like the Biobricks Foundation and iGEM. With our current understanding of transcriptional and translational regulation in cyanobacteria, we can redesign genetic regulatory elements and codon-optimised genes from distantly related organisms *in silico* and synthesize these constructs *in vitro* for expression in cyanobacteria. A team in France has developed codon-optimised (for expression in *E. coli, Synechocystis, Nostoc* and *Anabaena*) synthetic [Fe-Fe] hydrogenase genes from *Chlamydomonas reinhardtii* and *Clostridium acetobutylicum* that have been linked to a synthetic ferredoxin ligands derived from a chlamydomonal ferredoxin (Jaramillo, A., 2009, École Polytechnique, Palaiseau, France, privileged information). Recently, researchers in the Department of Photochemistry and Molecular Science at Uppsala University have characterised P_{trc} promoters, derived from the *lac*UV5 promoter, ribosomal binding sites and an expression vectors (pPMQAK1 and pPMQAC1) that are broadly functional in *E. coli* and in cyanobacteria (Brosius et al., 1985; Huang et al., 2010; Gibbons, 2010). With these tools, it is now possible to express prolific [Fe-Fe] hydrogenases and their respective maturation proteins in cyanobacteria. It is only a matter of time before viable hydrogen-yielding systems are in place.

Conclusion

In this post, I wanted to provide a glimpse of recent developments in photobiological hydrogen production (a field that I'm familiar with), but this is just one small piece of the puzzle. There are many teams of brilliant researchers around the globe chipping away at their own corners of sustainable energy. The future green economy will have to take a multi-faceted approach to energy, incorporating dozens of the most successful technologies, alongside behavioural and policy reform (i.e. less consumption, governmental regulation of greenhouse gasses, more localised food production, ecosystem preservation/ restoration etc.). My point was to inspire hope, despite all the doom and gloom surrounding climate change, by illuminating a subset of the work and ideas of our best and brightest, who toil day and night to build a better tomorrow. While we cannot ignore the perils that lie ahead, we must maintain a certain degree of hope in order to keep our heads above water and look towards a better horizon. Fear not, fellow humans - for every Senator James Inhofe, there are a hundred thousand scientists, social activists, community horticulturists and conservationists. By hook or by crook, we will shape our world into a more verdant and peaceful place (or die trying). We've got soul, but we're not soldiers.

Deep Ocean: Mother Nature's Very Own Science Fiction Movie

***Figure:** Deep Ocean: Mother Nature's Very Own Science Fiction Movie*

Contrary to popular belief (trekkies especially) space is not yet the final frontier - we still have plenty of unexplored frontier closer to home yet farther away from pop culture imagination: the oceans. The deep ocean remains one of the last truly explored regions, and it remains almost as mysterious as any distant galaxy.

According to Discovery Channel's epic documentary, Blue Planet, Seas of Life, "Over 60% of our planet is covered by water more than a mile deep. The deep sea is the largest habitat on earth and is largely unexplored. More people have travelled into space than have travelled to the deep ocean realm."

Our imaginations of the deep ocean verge on the alien, and as more information is uncovered, and more species discovered, we are shocked into amazement that the planet that we know so intimately, can support such alien life.

The deep ocean, including the benthic and abyssal zones makes up more than 80% of the ocean biome. These areas are cold and dark, being far below the reach of sunlight. Life is not as frequent as there are no plants or algae to feed on, and animals who call these regions home must find other means to live.

The deep ocean has markedly different physical characteristics from other parts of the Earth and even other parts of the ocean. Different pressures, temperatures, salinity, oxygen and other factors cause animals to adapt to these environments, developing unique tools of predation, diversion, and reproduction.

Light is one of the biggest factors that influence the direction of development in the deep ocean. There is no sunlight that reaches the waters of the deep ocean, and so this "twilight zone" is only lit by bioluminescent animals, which produce light using chemicals within their own bodies. Most animals need additional senses besides sight to navigate the deep ocean and survive either by catching prey or avoiding predators. Bioluminescent animals can use their light to attract prey, deflect predators, and communicate with other members of their species. This production of light can be the difference between being eaten, and escape, and also for securing your lunch.

One such predator is the deep sea angler fish. This predator uses a number of methods for effective survival in the deep ocean. Instead of wasting energy in vain pursuit of food, it sits and waits, preferring to lure and ambush its prey. Its lure is an elongated dorsal fin equipped with bioluminescent photophores at the tip of the fin. The small glowing portion of the fin draws in smaller fish in search of food, and then the fish is ambushed when within striking range of the angler's teeth. The angler fish's teeth are long, sharp and curved inward, to ensure no escape from prey. This fish more closely resembles a horror film than it does represent Mother Nature.

Ctenophores, commonly called Sea Gooseberries or Comb Jellies have a distinct pattern of eight rows (comb rows) of cilia, used for locomotion. The ciliated plates diffract light and cause a rainbow pattern of light to flash on the comb plates. Despite their mesmerizing beauty, the light attracts small prey which the Jellies consume whole.

Additionally many deep sea fish have large eyes, which help capture any residual light. This is the case with the strange *Micropinna microstoma*, which has tubular eyes shielded by a transparent head. The mechanism for the tubular eyes was recently discovered by researchers Bruce Robinson and Kim Reisenbichler, who found that the eyes can rotate within the transparent shield. The eyes can face upward when the fish is searching for food, and forward while feeding. Macropinna is one example of the development of the "barrel eye" whose tubular shape is ultrasensitive to capturing any remaining sunlight in the water. The transparent head, allows the eyes, while facing upward, to search for shadow, throwing

Light is not the only factor affecting life in the deep ocean, however. Pressure is another important influence. The pressure underwater increases 1 atmosphere for every 10 metres of depth. Pressures in the deep ocean can reach 1,000 atm. This makes studying the deep ocean difficult, since observation is limited and collected samples have to be pressurised to maintain stability. Animals in the deep ocean have adapted to the higher pressures through bodies without excess cavities that could collapse under intense pressure. Additionally, bones and flesh material are more gelatinous and soft which helps to withstand increased pressure. As a result, there are more jellyfish-like animals, worms and soft crustaceans than bony fish residing in the deep ocean.

One most amazing mollusc is the famed giant squid. Legends of the giant squid have been around since ancients have taken to the sea, and tales of a giant creature wraps its tentacles around ships and drags vessel and crew to the ocean depths. Although unlikely that any such creature is the cause of a sailor's watery grave, tales of the giant squid do have some merit. Specimens of the giant squid have been well documented, and samples collected around the world. But an even more amazing creature exists: the colossal squid. That's right, there is a squid that is bigger than giant. The Colossal squid, *Mesonychoteuthis hamiltonia,* lives in estimated depths of 2,000-2,200 metres. In 2007, a live colossal squid was brought to the surface near New Zealand, and was measured to be 10 metres long and weighed over 1,000 pounds. This half-ton behemoth is the largest invertebrate ever captured. Scientists identified it as a male of the species, and speculate that females could be even larger. Unlike giant squid, the colossal squid have swivelling hooks on each of the sucker disks at the ends of the feeding tentacles. These tools help this colossal predator wrangle with prey, and escape predators, most notably the Antarctic sperm whale.

Food sources are rare in the deep ocean, as the lack of sunlight and oxygen prevent algae and other plants to inhabit the depths. Life, however, seems to find new sustenance nearer hydrothermal vents. Dives near the Galapagos at depths of 2,700 km led to the discovery of a new ecosystem where life is sustained by the hot, mineral rich water which erupts from geysers in the seafloor. Heated by magma, the vents are found in or near mid-ocean ridges where new sea floor is formed. Metal sulfides are expelled along with other nutrients and communities of worms, and other animals thrive in the sunless environment. Most animals have special bacteria that turn hydrogen sulfide into food.

One such creature discovered near a hydrothermal vent is the Yeti Crab. This deep sea crab is blind, and is covered in long yellow hairs, or setae, which give the crab its name. This mythical crab was discovered near Easter Island, in the vents along the Pacific-Antarctic ridge by the research submarine *Alvin*. In an attempt to explain how creatures colonise the hydrothermal vents, specimens of the crab was collected and revealed that the crab's hairy limbs support large colonies of bacteria. This supports scientist's speculations that the crab in fact, farms the bacteria which live on the hydrogen sulfide possibly as a source of food. The hairs may also act as chemical and physical sensors to help find food or mates.

Although scientists have a good idea of what might be supporting this crab and many other creatures in the deep sea, much of it is still a mystery, as is the majority of how life works in the deep ocean. Although so close to home, the deep ocean remains virtually untouched, waiting for its secrets to be discovered

Hydrogen Storage Gets New Hope

A new method for "recycling" hydrogen-containing fuel materials could open the door to economically viable hydrogen-based vehicles. In an article appearing today in Angewandte Chemie, Los Alamos National Laboratory and University of Alabama researchers working within the U.S. Department of Energy's Chemical Hydrogen Storage Centre of Excellence describe a significant advance in hydrogen storage science.

Hydrogen is in many ways an ideal fuel for transportation. It is abundant and can be used to run a fuel cell, which is much more efficient than internal combustion engines. Its use in a fuel cell also eliminates the formation of gaseous byproducts that are detrimental to the environment.

For use in transportation, a fuel ideally should be lightweight to maintain overall fuel efficiency and pack a high energy content into a small volume. Unfortunately, under normal conditions, pure hydrogen has a low energy density per unit volume, presenting technical challenges for its use in vehicles capable of travelling 300 miles or more on a single fuel tank—a benchmark target set by DOE.

Consequently, until now, the universe's lightest element has been considered by some as a lightweight in terms of being a viable transportation fuel.

In order to overcome some of the energy density issues associated with pure hydrogen, work within the Chemical Hydrogen Storage Centre of Excellence has focused on using a class of materials known as chemical hydrides. Hydrogen can be released from these materials and potentially used to run a fuel cell. These compounds can be thought of as "chemical fuel tanks" because of their hydrogen storage capacity.

Ammonia borane is an attractive example of a chemical hydride because its hydrogen storage capacity approaches a whopping 20 percent by weight. The chief drawback of ammonia borane, however, has been the lack of energy-efficient methods to reintroduce hydrogen back into the spent fuel once it has been released. In other words, until recently, after hydrogen release, ammonia borane couldn't be adequately recycled.

Los Alamos researchers have been working with University of Alabama colleagues on developing methods for the efficient recycling of ammonia borane. The research team made a breakthrough when it discovered that a specific form of dehydrogenated fuel, called polyborazylene, could be recycled with relative ease using modest energy input. This development is a significant step toward using ammonia borane as a possible energy carrier for transportation purposes.

"This research represents a breakthrough in the field of hydrogen storage and has significant practical applications," said Dr. Gene Peterson, leader of the Chemistry Division at Los Alamos. "The chemistry is new and innovative, and the research team is to be commended on this excellent achievement." The Chemical Hydrogen Storage Centre of Excellence is one of three Centre efforts funded by DOE. The other two focus on hydrogen sorption technologies and storage in metal hydrides. The Centre of Excellence is a collaboration between Los Alamos, Pacific Northwest National Laboratory, and academic and industrial partners.

Referring to the work described in the *Angewandte Chemie* article, Los Alamos researcher John Gordon, corresponding author for the paper, stated, "Collaboration encouraged by our Centre model was responsible for this breakthrough. At the outset there were myriad potential reagents with which to attempt this chemistry."

"The predictive calculations carried out by University of Alabama professor Dave Dixon's group were crucial in guiding the experimental work of Los Alamos postdoctoral researcher Ben Davis," Gordon added. "The excellent synergy between these two groups clearly enabled this advance."

The research team currently is working with colleagues at The Dow Chemical Company, another Centre partner, to improve overall chemical efficiencies and move toward large-scale implementation of hydrogen-based fuels within the transportation sector.

Climate Sensitivity to CO2 More Limited than ExtremePprojection

A new study suggests that the rate of global warming from doubling of atmospheric carbon dioxide may be less than the most dire estimates of some previous studies and, in fact, may be less severe than projected by the Intergovernmental Panel on Climate Change report in 2007. Authors of the study, which was funded by the National Science Foundation's Paleoclimate Program and published online this week in the journal *Science*, say that global warming is real and that increases in atmospheric CO_2 will have multiple serious impacts.

However, the most Draconian projections of temperature increases from the doubling of CO_2 are unlikely.

"Many previous climate sensitivity studies have looked at the past only from 1850 through today, and not fully integrated paleoclimate date, especially on a global scale," said Andreas Schmittner, an Oregon State University researcher and lead author on the Science article. "When you reconstruct sea and land surface temperatures from the peak of the last Ice Age 21,000 years ago which is referred to as the Last Glacial Maximum and compare it with climate model simulations of that period, you get a much different picture. "If these paleoclimatic constraints apply to the future, as predicted by our model, the results imply less probability of extreme climatic change than previously thought," Schmittner added.

Scientists have struggled for years trying to quantify "climate sensitivity" which is how Earth will respond to projected increases of atmospheric carbon dioxide. The 2007 IPCC report estimated that the air near the surface of Earth would warm on average by 2 to 4.5 degrees (Celsius) with a doubling of atmospheric CO_2 from pre-industrial standards. The mean, or "expected value" increase in the IPCC estimates was 3.0 degrees; most climate model studies use the doubling of CO_2 as a basic index.

Some previous studies have claimed the impacts could be much more severe as much as 10 degrees or higher with a doubling of CO_2 although these projections come with an acknowledged low probability. Studies based on data going back only to 1850 are affected by large uncertainties in the effects of dust and other small particles in the air that reflect sunlight and can influence clouds, known as "aerosol forcing," or by the absorption of heat by the oceans, the researchers say.

To lower the degree of uncertainty, Schmittner and his colleagues used a climate model with more data and found that there are constraints that preclude very high levels of climate sensitivity.

The researchers compiled land and ocean surface temperature reconstructions from the Last Glacial Maximum and created a global map of those temperatures. During this time, atmospheric CO_2 was about a third less than before the Industrial Revolution, and levels of methane and nitrous oxide were much lower. Because much of the northern latitudes were covered in ice and snow, sea levels were lower, the climate was drier (less precipitation), and there was more dust in the air.

All these factor, which contributed to cooling Earth's surface, were included in their climate model simulations.

The new data changed the assessment of climate models in many ways, said Schmittner, an associate professor in OSU's College of Earth, Ocean, and Atmospheric Sciences. The researchers' reconstruction of temperatures has greater spatial coverage and showed less cooling during the Ice Age than most previous studies.

High sensitivity climate models more than 6 degrees suggest that the low levels of atmospheric CO_2 during the Last Glacial Maximum would result in a "runaway effect" that would have left Earth completely ice-covered.

"Clearly, that didn't happen," Schmittner said. "Though the Earth then was covered by much more ice and snow than it is today, the ice sheets didn't extend beyond latitudes of about 40 degrees, and the tropics and subtropics were largely ice-free except at high altitudes. These high-sensitivity models overestimate cooling."

On the other hand, models with low climate sensitivity less than 1.3 degrees underestimate the cooling almost everywhere at the Last Glacial Maximum, the researchers say. The closest match, with a much lower degree of uncertainty than most other studies, suggests climate sensitivity is about 2.4 degrees.

However, uncertainty levels may be underestimated because the model simulations did not take into account uncertainties arising from how cloud changes reflect sunlight, Schmittner said.

Reconstructing sea and land surface temperatures from 21,000 years ago is a complex task involving the examination of ices cores, bore holes, fossils of marine and terrestrial organisms, seafloor sediments and other factors. Sediment cores, for example, contain different biological assemblages found in different temperature regimes and can be used to infer past temperatures based on analogs in modern ocean conditions.

"When we first looked at the paleoclimatic data, I was struck by the small cooling of the ocean," Schmittner said. "On average, the ocean was only about two degrees (Celsius) cooler than it is today, yet the planet was completely different huge ice sheets over North America and northern Europe, more sea ice and snow, different vegetation, lower sea levels and more dust in the air.

"It shows that even very small changes in the ocean's surface temperature can have an enormous impact elsewhere, particularly over land areas at mid- to high-latitudes," he added.

Schmittner said continued unabated fossil fuel use could lead to similar warming of the sea surface as reconstruction shows happened between the Last Glacial Maximum and today.

"Hence, drastic changes over land can be expected," he said. "However, our study implies that we still have time to prevent that from happening, if we make a concerted effort to change course soon."

Other authors on the study include Peter Clark and Alan Mix of OSU; Nathan Urban, Princeton University; Jeremy Shakun, Harvard University; Natalie Mahowald, Cornell University; Patrick Bartlein, University of Oregon; and Antoni Rosell-Mele, University of Barcelona.

New Raptor Dinosaur Takes a Licking Keeps on Ticking

Raptor dinosaurs like the iconic Velociraptor from the movie franchise Jurassic Park are renowned for their "fear-factor." Their terrifying image has been popularised in part because members of this group possess a greatly enlarged talon on their foot analoguous to a butcher's hook. Yet the function of the highly recurved claw on the foot of raptor dinosaurs has largely remained a mystery to paleontologists. This week a collaboration of scientists unveil a new species of raptor dinosaur discovered in southern Utah that sheds new light on this and several other long-standing questions in paleontology, including how dinosaurs evolved on the "lost continent" of Laramidia (western North America) during the Late Cretaceous a period known as the zenith of dinosaur diversity. Their findings will be published in the journal *PLoS ONE.*

The new dinosaur dubbed Talos sampsoni is a member of a rare group of feathered, bird-like theropod dinosaurs whose evolution in North America has been a longstanding source of scientific debate, largely for lack of decent fossil material. Indeed, Talos represents the first definitive troodontid theropod to be named from the Late Cretaceous of North America in over 75 years. "Finding a decent specimen of this type of dinosaur in North America is like a lighting strike... it's a random event of thrilling proportions," said Lindsay Zanno, lead author of the study naming the new dinosaur. Zanno is an assistant professor of anatomy at the University of Wisconsin-Parkside and a research associate at the Field Museum of Natural History in Chicago, Illinois. Other members of the research team include Mike Knell (a graduate student at Montana State University) who discovered the new specimen in 2008 in the Kaiparowits Formation of Grand Staircase-Escalante National Monument (GSENM), southern Utah; Bureau of Land Management (BLM) paleontologist Alan Titus, leader of a decade-long paleontology reconnaissance effort in the monument; David Varricchio, Associate Professor of Paleontology, Montana State University; and Patrick O'Connor, Associate Professor of Anatomy, Ohio University Heritage College of Osteopathic Medicine.

Funding for the research was provided in part by the National Science Foundation, the Field Museum of Natural History, the Ohio

University Heritage College of Osteopathic Medicine, and the Bureau of Land Management. Zanno's research was supported by a John Caldwell-Meeker Fellowship and by a Bucksbaum Fellowship for young scientists. The bones of Talos sampsoni will be on exhibit for the first time in the Past Worlds Observatory at the new Utah Museum of Natural History, Salt Lake City, Utah.

The Nature of the Beast Troodontid theropods are a group of feathered dinosaurs closely related to birds. Members of this group are among the smallest non-avian dinosaurs known (as small as 100 grams) and are considered among the most intelligent. The group is known almost exclusively from Asia and prior to the discovery of Talos sampsoni, only two species were recognised in the Late Cretaceous of North America one of which, the infamous Troodon, was one of the first dinosaurs ever named from North America.

As a result of their distinctive teeth and the possible presence of seeds preserved as gut contents in one species, several scientists have proposed an omnivorous or herbivorous diet for at least some troodontids. Other species possess relatively blade-like teeth indicative of a carnivorous diet.

Zanno's own work on theropod diet suggests that extensive plant eating was confined to more primitive members of the group, with more advanced members of the clade like Troodon and Talos likely consuming at least some prey.

Several troodontid specimens have recently been discovered that not only support a close relationship with birds but also preserve remarkable evidence of bird-like behaviour.

These include extraordinary specimens such as eggs and embryos within nests that document transitional phases in the evolution of bird-like reproductive physiology and egg-laying behaviour, as well as specimens preserved in distinctive avian-like sleeping postures with their heads rotated back and tucked under their "wings." Other troodontids provide evidence of "four-winged" locomotor capabilities, and perhaps most extraordinary, plumage coloration.

With an estimated body mass of 38 kilograms, the newly discovered Talos sampsoni is neither the smallest nor largest troodontid known. Its skeleton indicates that the new species was much smaller and more slender than its famous cousin Troodon, which is known from sediments of the same age in the northern part of Laramidia (Alberta, Canada and Montana, USA). "Talos was fleet-footed and lightly built," Zanno says. "This little guy was a scrapper."

Interestingly, the holotype specimen of Talos also tells us something about theropod behaviour, particularly raptor behaviour. This is because the second toe that is, the one with the enlarged talon of the left foot of the new specimen is deformed, indicating that the animal suffered a fracture or bite during its life.

This Little Talos Takes a Beating

When the team first began studying the Talos specimen, they noticed some unusual features on the second digit of the left foot, but initially assumed they were related to the fact that it belonged to a new species. "When we realised we had evidence of an injury, the excitement was palpable," Zanno commented. "An injured specimen has a story to tell."

That's because evidence of injury relates to function. The manner in which an animal is hurt can tell you something about what it was doing during life. An injury to the foot of a raptor dinosaur, for example, provides new evidence about the potential function of that toe and claw. In order to learn about the injury to the animal's foot, the team scanned the individual bones using a high-resolution Computed Tomography (CT) scanner, similar to those used by physicians to examine bones and other organs inside the human body.

"Although we could see damage on the exterior of the bone, our microCT approach was essential for characterising the extent of the injury, and importantly, for allowing us to better constrain how long it had been between the time of injury and the time that this particular animal died," noted Patrick O'Connor, associate professor of anatomy at Ohio University.

After additional CT scanning of other parts of the foot, Zanno and her team realised that the injury was restricted to the toe with the enlarged claw, and the rest of the foot was not impacted. More detailed study suggested that the injured toe was either bitten or fractured and then suffered from a localised infection.

"People have speculated that the talon on the foot of raptor dinosaurs was used to capture prey, fight with other members of the same species, or defend the animal against attack. Our interpretation supports the idea that these animals regularly put this toe in harm's way," says Zanno.

Perhaps even more interesting is the fact that the injured toe exhibits evidence of bone remodelling thought to have taken place over a period of many weeks to months, suggesting that Talos lived with a serious injury to the foot for quite a long time. "It is clear from the bone remodelling that this animal lived for quite some time after the initial injury and subsequent infection, and that whatever it typically did with the enlarged talon on the left foot, whether that be acquire prey or interact with other members of the species, it must have been capable of doing so fairly well with the one on the right foot," added O'Connor.

Trackways made by animals closely related to Talos suggest that they held the enlarged talon off the ground when walking. "Our data support the idea that the talon of raptor dinosaurs was not used for purposes as mundane as walking," Zanno commented. "It was an instrument meant for inflicting damage."

What's in a Name?

The name Talos pays homage to a mythological Greek figure of the same name, believed to have protected the island of Crete by throwing stones at invading ships. It is said that the Greek Talos, who was often depicted as a winged bronze figure, could run at lightening speed and circled the ancient island three times a day. The dinosaur Talos belongs to a group of theropods known to have feathery integument (and in some cases "wings"), lived on the small island continent of Laramidia or west North America during the Late Cretaceous, and was also a fast runner.

The team chose the name Talos because of these similarities but also because the Greek Talos was said to have died from a wound to the ankle and it was clear that Talos had also suffered a serious wound to the foot. The species name "sampsoni" honours another famous figure Dr. Scott Sampson of the PBS series Dinosaur Train. Sampson, a research curator at the Utah Museum of Natural History and research faculty at the University of Utah, helped to spearhead a collaborative research effort known as the Kaiparowits Basin Project, a long-term research project that has been surveying and documenting the Late Cretaceous dinosaur fauna of the Kaiparowits Basin in southern Utah, with a focus on the Kaiparowits and Wahweap formations exposed in Grand Staircase-Escalante National Monument (GSENM).

Thus far this effort has resulted in the discovery of up to a dozen new dinosaurs from GSENM that are challenging previous ideas regarding Late Cretaceous dinosaur evolution and diversity within Laramidia and spurring new ideas regarding dinosaur biogeography in the region.

A Tale of Two Continents

Dinosaurs of the Late Cretaceous were living in a greenhouse world. A warm and equitable global climate that was devoid of polar ice caps and above average spreading at mid-oceanic ridges caused massive flooding of low-lying continental areas and created expansive epicontinental seaways. In North America, a shallow seaway running from the Gulf of Mexico through to the Arctic Ocean divided the continent into two landmasses, East America (Appalachia) and West America (Laramidia) for several million years during the Late Cretaceous. It was during this time that the dinosaurs achieved their greatest diversity, and scientists have been working overtime to understand why. Take for example the dinosaurs of Laramidia. The natural assumption is that being large bodied, those dinosaurs that lived on the small island continent would have roamed the whole area. However, recent fossil discoveries, particularly new dinosaurs from the Kaiparowits Formation, tell us that the true pattern is exactly the opposite.

Thus far the dinosaurs from the Kaiparowits Formation in southern Utah are entirely unique, even from those dinosaurs living just a few hundred miles to the north in what is now Montana and Alberta. Monument Paleontologist Alan Titus observed, "When we began looking in the remote Kaiparowits badlands we expected to see at least a few familiar faces. As it turns out, they are all new to science." And while recent discoveries from the Kaiparowits have substantiated this pattern for large-bodied herbivores like duck-bill and horned dinosaurs (for example Utahceratops), the pattern among small-bodied theropods was not clear. "We already knew that some of dinosaurs inhabiting southern Utah during the Late Cretaceous were unique," Zanno said, "but Talos tells us that the singularity of this ecosystem was not just restricted to one or two species. Rather, the whole area was like a lost world in and of itself."

A Monumental Discovery

Talos sampsoni is the newest member of a growing list of new dinosaur species that have been discovered in Grand Staircase Escalante National Monument (GSENM) in southern Utah. Former President Clinton founded the monument in 1996, in part to protect the world class paleontological resources entombed within its 1.9 million acres of unexplored territory. GSENM is one of the largest recently designated national monuments managed by the BLM, and one of the last pristine dinosaur graveyards in the US. The area has turned out to be a treasure trove of new dinosaur species, with at least 15 collected in just the past decade. Titus admits, "We had very few large fossils to substantiate the claim of 'World Class' paleontology when I started in 2000. Now, I feel GSENM could easily qualify as a world heritage site on the basis of its dinosaurs alone, dozens of which have been found preserving soft tissue." He also adds, "BLM support has been critical to the long term viability of the region's paleontology research and is paying off in countless ways both to the public and scientists."

Zanno, along with colleague Scott Sampson, named the first dinosaur from the monument Hagryphus giganteus in 2005. Hagryphus (widely touted in the press as the "turkey" dinosaur) is also a theropod dinosaur, but one that belongs to a different subgroup known as oviraptorosaurs (or egg thief reptiles). Other GSENM dinosaurs include five new horned dinosaurs including the recently described and bizarrely ornate Kosmoceratops and Utahceratops, three new duck-bill dinosaurs including the "toothy" Gryposaurus monumentensis, two new tyrannosaurs, as well as undescribed ankylosaurs (armoured dinosaurs), marine reptiles, giant crocodyliforms, turtles, plants, and a host of other organisms.

The discovery of a new troodontid from the monument is the latest in a long string of incredible fossil discoveries from the area. "I was surprised when I learned that I had found a new dinosaur," Knell said. "It is a rare discovery and I feel very lucky to be part of the exciting research happening here in the monument." Knell stumbled across the remains of Talos sampsoni while scouring the badlands of the Kaiparowits Formation for fossil turtles as part of his dissertation research.

Work continues every year in GSENM and new, significant fossil finds are made every field season. Considering there are hundreds of thousands of acres of outcrop that have yet to be surveyed, it is no exaggeration to claim the region will remain an exciting research frontier for decades to come.

Researchers Sequence Dark matter of Life

Researchers have developed a new method to sequence and analyse the dark matter of life the genomes of thousands of bacteria species previously beyond scientists' reach, from microorganisms that produce antibiotics and biofuels to microbes living in the human body. Scientists from UC San Diego, the J. Craig Venter Institute and Illumina Inc., published their findings in the Sept. 18 online issue of the journal *Nature Biotechnology*. The breakthrough will enable researchers to assemble virtually complete genomes from DNA extracted from a single bacterial cell. By contrast, traditional sequencing methods require at least a billion identical cells, grown in cultures in the lab. The study opens the door to the sequencing of bacteria that cannot be cultured the lion's share of bacterial species living on the planet.

"This part of life was completely inaccessible at the genomic level," said Pavel Pevzner, a computer science professor at the Jacobs School of Engineering at UC San Diego and a pioneer of algorithms for modern DNA sequencing technology.

Pevzner, in collaboration with UC San Diego mathematics professor Glenn Tesler and computer science postdoctoral researcher Hamidreza Chitsaz, developed an algorithm that dramatically improves the performance of software used to sequence DNA produced from a single bacterial cell. These programs traditionally recover 70 percent of genes.

"The new assembly algorithm captures 90 percent of genes from a single cell. Admittedly, it is not 100 percent. But it's almost as good as it gets for modern sequencing technologies: today biologists typically capture 95 percent of genes but they need to grow a billion cells to accomplish it," said Tesler.

Bacteria play a vital role in human health. They make up about 10 percent of the weight of the human body and can be found anywhere from the stomach to the mouth. Some, like *E. coli*, can wreak havoc. Others help us digest. Yet others, recent studies have found, can change the way we behave by, for example, tricking us into eating more than we need. That's why it is crucial to analyse bacteria's genomes, which in turn help scientists understand bacteria's behaviour.

Modern sequencing machines require DNA from one billion bacterial cells to produce a complete genome. Biologists usually grow the required amount of bacteria in cultures in the lab. That is how they obtained enough DNA to sequence *E. coli*. But a wide majority of bacteria 99.9 percent according to some estimates cannot be cultured in the lab because they live in specific conditions and environments that are hard to reproduce, for example in symbiosis with other bacteria or on an animal's skin.

Enter Multiple Displacement Amplification (MDA) technology, developed about a decade ago by Professor Roger Lasken, now at the

Venter Institute and co-author of the Nature Biotechnology study. MDA can be used on bacteria that can't be cultured in the lab. The technology is the equivalent of a copy machine that starts from a single cell and makes copies of fragments of its genome until it produces the equivalent of one billion cells. In 2005, Lasken and colleagues used MDA to sequence DNA produced from a single cell for the first time with funding from the Department of Energy. However, while MDA is an ingenious cellular copy machine, it gives sequencing software programs a hard time. The DNA copies that MDA makes carry various errors and are not amplified uniformly: some pieces of the genome are copied thousands of times, and others only once or twice. Modern sequencing algorithms aren't equipped to deal with these disparities. In fact, they tend to discard bits of the genome that were replicated only a few times as sequencing errors, even though they could be key to sequencing the whole genome. The algorithm developed by Pevzner's team changes that. It retains these genome pieces and uses them to improve sequencing.

Researchers sequenced a single cell of *E. coli* with this method to verify the accuracy of the algorithm and recovered 91 percent of its genes, doing nearly as well as conventional sequencing from cultured cells. This provides enough data to answer many important biological questions, such as what antibiotics a species of bacteria produces. It also, for the first time, enables researchers to perform in-depth studies to figure out which proteins and peptides the bacteria living in human beings use to communicate with each other and with their host.

The scientists then turned to a species of marine bacteria that had never been sequenced before part of the dark matter of life. They not only sequenced its genome, but also analysed it and were able to get information about how it lives and moves. The fairly complete and annotated genome they obtained was the first genome obtained via MDA to be deposited in GenBank, the genetic sequence database at the National Institutes of Health. With the help of the new algorithm developed by Pevzner and colleagues, thousands more are set to follow. Pevzner's team is at work on a second-generation version of the algorithm. Lasken and his team plan to continue their work on improving MDA as well.

Lasken keeps a few hundred tubes filled with unsequenced bacteria in his laboratory at the Venter Institute in La Jolla, Calif. Each represents a bacterial terra incognita that scientists soon will explore using the method developed through the combined efforts of researchers at the UC San Diego Jacobs School of Engineering, the Venter Institute and Illumina.

"It's a very big step forward," Lasken said. The research was partially supported by grants from the National Human Genome Research Institute and the Alfred P. Sloan Foundation and by a grant from the National Institutes of Health.

Hitchhiking Snails Fly from Ocean to Ocean

Smithsonian scientists and colleagues report that snails successfully crossed Central America, long considered an impenetrable barrier to marine organisms, twice in the past million years both times probably by flying across Mexico, stuck to the legs or riding on the bellies of shorebirds and introducing new genes that contribute to the marine biodiversity on each coast. "Just as people use aeroplanes to fly overseas, marine snails may use birds to fly over land," said Mark Torchin, staff scientist at the Smithsonian. "It just happens much less frequently. There's also a big difference between one or two individuals ending up in a new place, and a really successful invasion, in which several animals survive, reproduce and establish new populations."

The discovery of the hitchhiking snails, published in *Proceedings of the Royal Society: B*, has broad implications. "Not only snails, but many intertidal organisms may be able to 'fly' with birds," said first author of the study, Osamu Miura, assistant professor at Japan's Kochi University and former postdoctoral fellow at the Smithsonian Tropical Research Institute in Panama.

Chance events that occur only once in a great while may be extremely important in the history of life. In 1940, George Gaylord Simpson, who studied natural history as recorded in fossils, coined the term "sweepstakes dispersal" to describe the unlikely events in which animals cross over a barrier resulting in major consequences for the diversity of life on Earth. Simpson was thinking about land-based animals that might "get lucky" and cross between continents or islands by floating on rafts of debris. Sometimes such events result in devastating biological invasions introducing new diseases, wiping out resident species or causing economic damage to food crops.

The idea of land snails hitching rides on birds goes back to Charles Darwin, who speculated that migratory birds could transport snails to distant places. In fact, birds are thought to have carried land snails 5,500 miles from Europe to Tristan de Cunha Island in the

South Atlantic Ocean and back. But this is the first report of a marine snail "flying" from one ocean to another.

Scientists working at the Smithsonian in Panama have long been interested in how the rise of the Central American land bridge more than 3 million years ago drove speciation and increased biodiversity. It formed a barrier between marine species, some of which evolved in their new surroundings, becoming new "sister" species that could no longer mate with their former relatives. By studying the genetics of two sister species of Horn Snails, Cerithideopsis californica and C. pliculosa, collected at 29 different locations in mudflats and mangrove habitats from California to Panama on the Pacific and from Texas to Panama on the Atlantic, the researchers discovered that, about 750,000 years ago, these snails invaded the Atlantic from the Pacific, and then, about 72,000 years ago, Atlantic populations returned to invade Pacific shores.

"Shorebirds mostly move back and forth across Central America via a couple of flyways," said Torchin. "We think that the snails were able to cross the Isthmus of Tehuantepec in Mexico because it's a major bird flyway and is a relatively flat and narrow stretch of land with ideal tidal flat habitat on either side."

"There is a chance that the hitchhiking snails benefited native populations by bringing in new genes that helped them resist common parasites that castrate the snails and keep them from reproducing," said Ryan Hechinger, associate research biologist at the University of California, Santa Barbara. "Now we are looking at the parasite genes to see if they jumped Central America too."

Understanding that such hitchhiking occurs can help reveal where new species might have become established or where they might establish in the future," said Eldredge Bermingham, STRI director and staff scientist." I am here in Panama watching as snails fly over my head. Tongue in cheek, I fail to understand why others did not notice this before! I suspect our interpretation of this phylogeographic pattern would make George Gaylord Simpson smile."

Honduran Earthquake of 2009 Destroyed Half of Coral Reefs of Belizean Barrier Reef Lagoon

Earth's coral reefs have not been faring well in recent decades, facing multiple threats from pollution, disease, elevated water temperatures, and overfishing. Often referred to as the "rainforests of the Sea," coral reefs support a wide variety of marine life, help protect shorelines, and contribute significantly to tourism and the fishing industry. A new study looks at a rare but catastrophic impact on reefs: the damage caused by natural disasters such as an earthquakes. In May of 2009, a powerful, magnitude-7.3 earthquake shook the western Caribbean, causing lagoonal reefs in Belize, 213 kilometres (132 miles) from the epicentre, to avalanche and slide into deeper water. As reported in a preprint article of *Ecology*, a journal of the Ecological Society of America, Richard Aronson of the Florida Institute of Technology and colleagues analysed data that suggest how the history of the reef will influence its recovery.

During the quarter-century before the earthquake struck, the reefs had gone through mass mortalities of two sequentially dominant coral species. Novel events in their own right, these mass mortalities were instantly "rendered moot" on half the reefs, which were destroyed when the earthquake hit.

Aronson and colleagues' work focused on a 375 square kilometre (144 square mile) area of the Belizean Barrier Reef, which they monitored from 1986 to 2009. The group revisited 21 sites in 2010 to determine the impacts of the earthquake. They found that approximately half the reef slopes had slabbed off and slid into deeper water. Only sediment and the skeletal debris of corals remained.

Beginning in 1986, a bacterial infection called white-band disease killed virtually all the then-dominant staghorn coral (*Acropora cervicornis*) in the study area. By 1995, lettuce coral (Agaricia tenuifolia) had taken over the number-one spot. But when high temperatures from the 1998 El Nino-Southern Oscillation, which were aggravated by global climate change, caused mass coral bleaching, lettuce coral disappeared. An encrusting sponge (*Chondrilla caribensis*) colonised its skeletal remains, along with seaweed. What's astonishing about this series of events, say the authors, is that as evidenced by radiocarbon-dating of reef cores staghorn coral had dominated the reefs for nearly 4,000 years.

"The prior losses of both staghorn and lettuce corals drastically weakened the resilience of the coral assemblages on the reef slopes," says lead author Aronson. "In other words, if neither white-band disease nor bleaching had occurred, staghorn coral might have continued its millennial-scale dominance of the areas not destroyed by the quake."

The authors project that recovery to a coral-dominated state is unlikely in the near future, because corals in the undamaged areas had been killed previously. The situation is unlikely to change unless the way we manage reef resources improves dramatically.

Marine protected areas are meant to sustain an area's ecological, cultural, and economic benefits for future generations. Yet creating and managing these areas is easier said than done. Aronson and colleagues contend that extreme events, such as earthquakes, lava flows, and tsunamis, should be taken into account when determining the size of and managing such protected areas. "The rhetoric of conservation often includes the appeal of preserving ecosystems so that our children's children can enjoy Nature's bounty," says Aronson. "That translates to about 200 years, but ecosystems last far longer than three generations of their human stewards. We challenge marine conservationists to plan on a millennial scale. Rare, catastrophic events are the backdrop to human actions. Those rare events should be factored into determining the sizes of marine reserves and their levels of protection, whatever else might be expected to happen along the way. After all, a once in a thousand year disaster could still occur next week.

Ongoing Global Biodiversity Loss Unstoppable with Protected Areas Alone: Study

Continued reliance on a strategy of setting aside land and marine territories as "protected areas" is insufficient to stem global biodiversity loss, according to a comprehensive assessment published today in the journal *Marine Ecology Progress Series.* Despite impressively rapid growth of protected land and marine areas worldwide - today totalling over 100,000 in number and covering 17 million square kilometres of land and 2 million square kilometres of oceans - biodiversity is in steep decline.

Expected scenarios of human population growth and consumption levels indicate that cumulative human demands will impose an unsustainable toll on the Earth's ecological resources and services accelerating the rate at which biodiversity is being loss.

Current and future human requirements will also exacerbate the challenge of effectively implementing protected areas while suggesting that effective biodiversity conservation requires new approaches that address underlying causes of biodiversity loss - including the growth of both human population and resource consumption.

Says lead author Camilo Mora of University of Hawaii at Manoa: "Biodiversity is humanity's life-support system, delivering everything from food, to clean water and air, to recreation and tourism, to novel chemicals that drive our advanced civilization. Yet there is an increasingly well-documented global trend in biodiversity loss, triggered by a host of human activities."

"Ongoing biodiversity loss and its consequences for humanity's welfare are of great concern and have prompted strong calls for expanding the use of protected areas as a remedy," says fellow author Peter F. Sale, Assistant Director of the United Nations University's Canadian-based Institute for Water, Environment and Health.

"While many protected areas have helped preserve some species at local scales, promotion of this strategy as a global solution to biodiversity loss, and the advocacy of protection for specific proportions of habitats, have occurred without adequate assessment of their potential effectiveness in achieving the goal."

Drs. Mora and Sale warn that long-term failure of the protected areas strategy could erode public and political support for biodiversity conservation and that the disproportionate allocation of available resources and human capital into this strategy precludes the development of more effective approaches.

The authors based their study on existing literature and global data on human threats and biodiversity loss.

"The global network of protected areas is a major achievement, and the pace at which it has been achieved is impressive," says Dr. Sale. "Protected areas are very useful conservation tools, but unfortunately, the steep continuing rate of biodiversity loss signals the need to reassess our heavy reliance on this strategy."

The study says continuing heavy reliance on the protected areas strategy has five key technical and practical limitations:

- Expected growth in protected area coverage is too slow: While over 100,000 areas are now protected worldwide, strict enforcement occurs on just 5.8% of land and 0.08% of ocean. At current rates, it will take between 185 years in the case of land and 80 years for oceans to cover 30% of the world's ecosystems with protected areas - a minimum target widely advocated for effective biodiversity conservation.

This slow pace contrasts sharply with the rapid growth of threats, including climate change, habitat loss and resource exploitation, predicted to cause the extinction of many species even before 2050.

- The size and connectivity of protected areas are inadequate : To ensure species' survival, protected areas must be sufficiently large to sustain viable populations in the face of the inevitable mortality of some individuals trespassing their borders, and areas must be close enough together for a healthy exchange of individuals among protected populations. Globally, however, over 30% of the protected areas in the ocean, and 60% on land are smaller than 1 square kilometer too small for many larger species. And they tend to be too far apart to allow a sufficient exchange among populations for most species.
- Protected areas only ameliorate certain human threats: Biodiversity loss is triggered by a host of human stressors including habitat loss, overexploitation, climate change, pollution and invasive species. Yet protected areas are useful primarily against overexploitation and habitat loss. Since the remaining stressors are just as deleterious, biodiversity can be expected to continue declining as it has done until now. The study shows that approximately 83% of protected areas on the sea and 95% of protected areas on land are located in areas with continuing high impact from multiple human stressors.
- Underfunding: Global expenditures on protected areas today are estimated at US $6 billion per year and many areas are insufficiently funded for effective management. Effectively managing existing protected areas requires an estimated $24 billion per year - four times current expenditure. Despite strong advocacy for protected areas, budget growth has been slow and it seems unlikely that it will be possible to raise funding appropriate for effective management as well as for creation of the additional protected areas as is advocated.
- Conflicts with human development: Humanity's footprint on Earth is ever expanding in efforts to meet basic needs like housing and food. If it did prove possible to place the recommended 30% of world habitats under protection, intense conflicts with competing human interests are inevitable many people would be displaced and livelihoods impaired. Forcing a trade-off between human development and sustaining biodiversity is unlikely to lead to a solution with biodiversity preserved.

Concludes Dr. Mora: "Given the considerable effort and widespread support for the creation of protected areas over the past 30 years, we were surprised to find so much evidence for their failure to effectively address the global problem of biodiversity loss. Clearly, the biodiversity loss problem has been underestimated and the ability of protected areas to solve this problem overestimated."

The authors underline the correlations between growing world population, natural resources consumption and biodiversity loss to suggest that biodiversity loss is unlikely to be stemmed without directly addressing the ecological footprint of humanity. Based upon previous research, the study shows that under current conditions of human comsumption and conservative scenarios of human population growth, the cummulative use of natural resources of humanity will amount to the productivity of up to 27 Earths by 2050.

Protected areas are a valuable tool in the fight to preserve biodiversity. We need them to be well managed, and we need more of them, but they alone cannot solve our biodiversity problems," adds Dr. Mora. "We need to recognise this limitation promptly and to allocate more time and effort to the complicated issue of human overpopulation and consumption."

"Our study shows that the international community is faced with a choice between two paths," Dr. Sale says. "One option is to continue a narrow focus on creating more protected areas with little evidence that they curtail biodiversity loss. That path will fail. The other path requires that we get serious about addressing the growth in size and consumption rate of our global population."

Census of Marine Life Publishes Historic Roll Call of Species in 25 Key World Areas

Representing the most comprehensive and authoritative answer yet to one of humanity's most ancient questions "what lives in the sea?" Census of Marine Life scientists today released an inventory of species distribution and diversity in key global ocean areas. Scientists combined information collected over centuries with data obtained during the decade-long Census to create a roll call of species in 25 biologically representative regions from the Antarctic through temperate and tropical seas to the Arctic.

Their papers help set a baseline for measuring changes that humanity and nature will cause.

Published by the open access journal *PLoS ONE*, the landmark collection of papers and overview synthesis will help guide future decisions on exploration of still poorly-explored waters, especially the abyssal depths, and provides a baseline for still thinly-studied forms, especially small animals.

Australian and Japanese waters, which each feature almost 33,000 forms of life that have earned the status of "species" (and thus a scientific name such as *Carcharodon carcharias*, a.k.a. the great white shark), are by far the most biodiverse. The oceans off China, the Mediterranean Sea and the Gulf of Mexico round out the top five areas most diverse in known species. In a prelude to the ultimate summary of the landmark, decade-long marine census, to be released Oct. 4 in London, national and regional committees of the Census compiled the inventory of known and new species in the 25 key marine regions.

The 13 committees include over 360 scientists whose collective knowledge, including published and unpublished data, was assembled to create the initial profile of known marine biodiversity in Antarctica, Atlantic Europe, Australia, Baltic Sea, Brazil, Canada (East, West and Arctic), Caribbean Sea, China, Indian Ocean, Japan, Mediterranean Sea, New Zealand, South Africa, South America (Tropical East Pacific and Tropical West Atlantic), South Korea, the Humboldt Current, the Patagonian Shelf, and the USA (Northeast, Southeast, Hawaii, Gulf of Mexico, and California).

Major inventories continue in highly diverse areas such as Indonesia, Madagascar and the Arabian Sea, which have yet to report.

Scientists find that the number of known, named species contained in the 25 areas ranged from 2,600 to 33,000 and averaged about 10,750, which fall into a dozen groups. On average, about one-fifth of all species were crustaceans which, with mollusks and fish, make up half of all known species on average across the regions. The full breakdown follows:

- 19% Crustaceans (including crabs, lobsters, crayfish, shrimp, krill and barnacles),
- 17% Mollusca (including squid, octopus, clams, snails and slugs)
- 12% Pisces (fish, including sharks)
- 10% Protozoa (unicellular micro-organisms)
- 10% algae and other plant-like organisms
- 7% Annelida (segmented worms)
- 5% Cnidaria (including sea anemones, corals and jellyfish)
- 3% Platyhelminthes (including flatworms)

- 3% Echinodermata (including starfish, brittle stars, sea urchins, sand dollars and sea cucumbers)
- 3% Porifera (including sponges)
- 2% Bryozoa (mat or 'moss animals')
- 1% Tunicata (including sea squirts) .

The rest are other invertebrates (5%) and other vertebrates (2%). The scarce 2% of species in the "other vertebrates" category includes whales, sea lions, seals, sea birds, turtles and walruses. Thus some of the best-known marine animals comprise a tiny part of marine biodiversity.

The authors note that their work constitutes a roll call of marine plant and animal species – either present or unknown in 25 regions. It does not represent their abundance or biomass.

The Most Cosmopolitan Species

Many species appear in more than one region. Current holders of the title "most cosmopolitan" marine species are two opposite kinds: microscopic plants (algae) and single-celled animals called protozoa and copepod in the plankton, and the seabirds and marine mammals that traverse the oceans throughout their lives.

Among fish, the manylight viperfish) can be considered the Everyman of the deep ocean. Census data shows the fish has been recorded in more than one-quarter of the world's marine waters.

How the microscopic species can be cosmopolitan is still a subject of research, and may be due to their ability to survive unsuitable environmental conditions and then reach enormous abundance in a suitable environment.

Says Patricia Miloslavich of Universidad Simón Bolívar, Venezuela, co-senior scientist of the Census and leader of the regional studies: "To create this baseline, the Census of Marine Life explored new areas and new ecosystems, discovering new species and records of species in new places.

"We reviewed what had been documented through the huge efforts of scientists in years past. However, most of this information was scattered or unavailable except at a very local level. The Census has made a tremendous contribution by bringing order to chaos. This previously scattered information is now all reviewed, analysed and presented in a collection of papers at an open access journal."

Says lead author of the summary, Mark Costello of the Leigh Marine Laboratory, University of Auckland, New Zealand: "Sparse, uneven marine sampling in much of the world underlies this initial inventory, and future research will undoubtedly alter the profile presented today."

He adds that finding such great difference in the proportions of species across regions challenges assumptions that scientists can extrapolate knowledge of biodiversity from one location to another.

"This inventory was urgently needed for two reasons," says Dr. Costello. "First, dwindling expertise in taxonomy impairs society's ability to discover and describe new species. And secondly, marine species have suffered major declines in some cases 90% losses due to human activities and may be heading for extinction, as happened to many species on land."

Regional results:

- Even less diverse regions such as the Baltic or Northeast USA still have about 4,000 known species.
- Relative to its volume of water, the Baltic, followed by China, has some of the highest known diversity.
- Relative to their seabed area, South Korea, China, South Africa and the Baltic, had most species.
- The relative contribution of different kinds of life to the species in each region varied greatly and enigmatically. While variation in research effort may be part of the explanation, it also seems that species have not flourished equally around the world.
- Crustaceans (including crabs, lobsters, crayfish, shrimp, krill and barnacles) contributed 22% to 35% of species for Alaska, Antarctica, Arctic, Brazil, California, Caribbean, and Humboldt regions, but only 10% for the Baltic.
- Mollusks (clams, snails, squid and slugs) contributed 26% of the species in Australia and Japan, but only 5% to 7% of the species in the Baltic, California, Arctic, and eastern and western Canada.
- Fish comprised 28% of species in the Tropical West Atlantic and Southeast USA, but only 3% to 6% for the Arctic, Antarctica, Baltic, and Mediterranean;
- Of the less species-rich groups, Annelida (worms) contributed 28% of the species for the Tropical Eastern Pacific, but only 3% for Japan.

- Plants and algae (mostly algae) contributed about one third of species in the Baltic, Arctic, Atlantic Europe, and Western Canada, but few in Antarctica, Caribbean, China, Humboldt, Tropical Eastern Pacific, and Tropical Western Atlantic.

Where to Find Unique, "Endemic" or Alien Invasive Species

- The number of unique "endemic" species seen nowhere else on Earth provides another measure of biodiversity. The relatively isolated regions Australia, New Zealand, Antarctica and South Africa have the most endemic species. They may have suffered fewer extinctions from climate cooling thousands of years ago during glaciation. Or, species from regions that escaped glaciers may have reached them more easily when the glaciers melted.
- Endemics comprise about half of New Zealand and Antarctic marine species and a quarter of those in Australian and South Africa. The waters of the Caribbean, China, Japan, and Mediterranean each have less than 2,000 endemic species, and the Baltic only 1 – a seaweed (*Fucus radicans)*
- To encounter invasive species, visit the Mediterranean. It had the most alien species among the 25 regions with over 600 (4% of the all species inventoried), most of which arrived from the Red Sea via the Suez Canal.
- Many aliens have also invaded the European Atlantic, New Zealand, Australian Pacific, and Baltic waters. Mollusks, crustaceans, and fish were the most common invading aliens.

Says Dr. Ian Poiner, CEO of the Australian Institute for Marine Science and Chair of the Census Scientific Steering Committee: "Consider that a well-informed person walking along a familiar seashore might identify 20 species or so; a fish monger perhaps 100. Even in the world's least diverse marine regions, there are 50 to 100 times as many named species than an expert would know without resorting to field guides."

Many of the species records used for the report are part of the 10 year old Ocean Biogeographic Information System (OBIS), a massive global database of what/where records and a major Census legacy. An interested person can find precise places on a world map where a marine organism has been reliably observed.

OBIS has consolidated almost 30 million records from the Census projects and more than 800 databases contributed by institutions around the world." Says Edward Vanden Berghe, who leads development and management of the database: "A map of records in OBIS today underlines the uneven sampling of oceanic regions," he adds. "So even as records accumulate, the importance of orderly sampling grows." Almost all the species in the key regional areas are included in the unprecedented list of 185,000 marine species created by the World Register of Marine Species (WoRMS), an affiliate of the Census of Marine Life.

How Much is Unknown?

In October, the Census will release its latest estimate of all marine species known to science, including those still to be added to WoRMS and OBIS. This is likely to exceed 230,000.

According to a recent open access Census of Marine Life paper in *Zootaxa* by US expert Bill Eschmeyer and colleagues, the number of marine fish species in mid-February stood at 16,764, and was growing at a rate of 100 to 150 per year. They estimate about 5,000 marine fish species have yet to be discovered and described – twice the number described in the last 19 years – for a projected total of approximately 21,800 marine fish species around the world.

And for every marine species of all kinds known to science, Census scientists estimate that at least four have yet to be discovered. In a few taxonomic groups, like fish, scientists believe more than 70% of species have been discovered, but for most other groups likely less than one-third are known. Scientists believe that the tropics, deep-seas and southern hemisphere hold the most undiscovered marine species.

The proportion of species not yet described is estimated at 39 to 58% in Antarctica, 38% for South Africa, 70% for Japan, 75% for the Mediterranean deep-sea, and more than 80% for Australia.

New Zealand has more than 4,100 undescribed species in its specimen collections, which would comprise 25% of the country's known marine species, but clearly is a minimum estimate because many species have not been collected and distinguished in collections.

Citizens of the Sea

"At the end of the Census of Marine Life, most ocean organisms still remain nameless and their numbers unknown," says renowned biologist Nancy Knowlton of the Smithsonian Institution, leader of the Census' coral reef project and author of a new book published by the

National Geographic Society for release September 14. "Citizens of the Sea: Wondrous Creatures from the Census of Marine Life" is one of three books marking the Census' conclusion.

"This is not an admission of failure. The ocean is simply so vast that, after 10 years of hard work, we still have only snapshots, though sometimes detailed, of what the sea contains. But it is an important and impressive start."

Dr. Knowlton's book, written in everyday language and populated with scores of images, draws on discoveries of Census scientists and their colleagues, past and present. It chronicles "the variety, beauty, weirdness and wonder that characterises life in the sea."

"The sea today is in trouble," says Dr. Knowlton. "Its citizens have no vote in any national or international body, but they are suffering and need to be heard. Much has changed just in the few decades that I have spent on and under the sea, but it remains a wondrous and enriching place, and with care it can become even more so."

Greatest Threats

According to the Census studies published in *PLoS ONE*, the main threats to marine life to date have been overfishing, lost habitat, invasive species and pollution, although the relative importance of the threats varied among regions. Emerging threats include rising water temperature and acidification, and the enlargement of areas characterised by low oxygen content (called hypoxia) of seawater. These too will vary regionally (surface temperature, for example) whereas others are more global (such as acidification).

Overfishing not only depletes the exploited fish themselves but also depletes other species like turtles, albatrosses, sharks and mammals, caught unintentionally. It alters food webs within ecosystems.

Coastal urbanisation, sediment runoff and nutrients in sewage and fertilizer washed from the land and causing eutrophication and hypoxia are destroying marine habitats.

The more enclosed seas Mediterranean, Gulf of Mexico, China's shelves, Baltic, and Caribbean were reported to have the most threatened biodiversity.

State-of-knowledge Index

Census scientists created a relative "state-of-knowledge index," grading each region according to how well it was known, including the availability of guides to the identification of species and their number of taxonomic experts.

In a nutshell, the studies found that while the depth of knowledge varies across regions, knowledge in all regions is inadequate.

Australia, China and all three European regions scored the highest index results while the Tropical West Atlantic, Tropical East Pacific and Canadian Arctic were well below average. But, even in regions with the highest index scores, knowledge of marine biodiversity is poor. In Australia it is estimated that only about 10% of marine life in its Exclusive Economic Zone is known. Scientists say the availability of comprehensive species identification guides strongly boost the discovery and management of marine biodiversity resources.

"We must increase our knowledge of unknown biodiversity more quickly, lest much of it is lost without even being discovered," says Dr. Miloslavich. "International sharing of data, expertise and resources, as has been accomplished through the Census of Marine Life, is the most cost-effective way of achieving this."

The Census papers collection, freely available Aug. 2 at *PLoS ONE*, includes links to maps, databases and a suite of the first nine regional papers on which the summary drew, with several more to be added in weeks to come.

And there are Census reports on several more regions anticipated in years to come. An exploration currently underway in the species-rich Timor and Arafura Seas, facilitated by Dr. Antonio (Tonny) Wagey, leader of the Census' National Committee in Indonesia, will enrich the Indonesian report. Another this past spring, led by Philippe Bouchet of the Muséum National d'Histoire Naturelle, Paris, discovered a vast array of marine life in the Deep South of Madagascar.

4

Ocean Resources

Deep-sea Exploration

Deep-sea exploration is the investigation of physical, chemical, and biological conditions on the sea bed, for scientific or commercial purposes. Deep-sea exploration is considered as a relatively recent human activity compared to the other areas of geophysical research, as the depths of the sea have been investigated only during comparatively recent years. The ocean depths still remain as a largely unexplored part of the planet, and form a relatively undiscovered domain. In general, modern scientific Deep-sea exploration can be said to have begun when French scientist Pierre Simon de Laplace investigated the average depth of the Atlantic ocean by observing tidal motions registered on Brazilian and African coasts. He calculated the depth to be 3,962 m (13,000 ft), a value later proven quite accurate by soundings measurement.

Later on, with increasing demand for submarine cables installment, accurate soundings was required and the first investigations of the sea bottom were undertaken. First deep-sea life forms were discovered in 1864 when Norwegian researchers obtained a sample of a stalked crinoid at a depth of 3,109 m (10,200 ft). The British Government sent out the Challenger expedition (a ship called the HMS Challenger) in 1872 which discovered 715 new genera and 4,417 new species of marine organisms over the space of 4 years. The first instrument used for deep-sea investigation was the sounding weight, used by British explorer Sir James Clark Ross. With this instrument, he reached a depth of 3,700 m (12,140 ft) in 1840. The Challenger expedition used similar instruments called Baillie sounding machines to extract samples from the sea bed.

In 1960, Jacques Piccard and US Navy Lieutenant Donald Walsh descended in Trieste to make the deepest dive in history: 10,915 metres (35,810 ft).

Brief History

Throughout history, scientists have relied on a number of instruments to measure, map, and observe the ocean's depths. One of the first instruments used to examine the seafloor was the sounding weight. Ancient Viking sailors took measurements of sea depth and sampled seafloor sediments with this instrument, which consisted of a lead weight with a hollow bottom attached to a line. Once the weight reached the sea bottom and collected a sample of the seabed, the line was hauled back on board ship and measured in fathom. Cornelius Drebbel, a Dutch architect, is generally given credit for construction of the first submarine. His submersible boat consisted of a wooden frame sheathed in animal skin. Oars, with its openings were sealed with tight-fitting leather flaps, extended out the sides to propel the craft through the water, at depths up to 4.6 metres (15 ft). Drebel tested his submarine in the Thames River in England in sometime between 1620 and 1624. It is believed that King James I may have enjoyed a short ride in the craft.

However, the nature of the deep ocean remained an unrevealed mystery until the mid-19th century. Scientists and artists alike imagined the deep sea as a lifeless soup of placid water, French author Jules Verne who helped pioneer the science-fiction genre portrayed the deep ocean as contained in a bowl of static rock in his "Twenty Thousand Leagues under the Sea". By the late 1860s, controversial modern scientific theories, the origin of life by evolution and the enormity of geologic time had created a foundation of scientific curiosity and provoked a rising interest in marine exploration. The Royal Society of England thus initiated an ambitious oceanographic mission to expand a scarce collection of existing marine data that included Charles Darwin's observations during the voyage of the HMS Beagle (1831–1836), a bathymetric chart created by U.S. Navy Lt. Matthew Maury to aid installation of the first transcontinent telegraph cables in 1858, and a few examples of deep marine creatures.

From 1872 to 1876, a landmark ocean study was carried out by British scientists aboard HMS Challenger, a sailing vessel that was redesigned into a laboratory ship. The HMS Challenger expedition covered 127,653 km (68,890 nautical miles), and shipboard scientists collected hundreds of samples, hydrographic measurements, and specimens of marine life. They are also credited with providing the first real view of major seafloor features such as the deep ocean basins. They discovered more than 4,700 new species of marine life, including deep-sea organisms.

Deep-sea exploration advanced considerably in the 1900s thanks to a series of technological inventions, ranging from sonar system to detect the presence of objects underwater through the use of sound to manned deep-diving submersibles such as DSV Alvin. Operated by the Woods Hole Oceanographic Institution, Alvin is designed to carry a crew of three people to depths of 4,000 metres (13,124 ft). The submarine is equipped with lights, cameras, computers, and highly maneuverable robotic arms for collecting samples in the darkness of the ocean's depths. However, the voyage to the ocean bottom is still a challenging experience. Scientists are working to find ways to study this extreme environment from the shipboard. With more sophisticated use of fibre optics, satellites, and remote-control robots, scientists one day may explore the deep sea from a computer screen on the deck rather than out of a porthole.

Milestones of Early Deep Sea Exploration

The extreme conditions in the deep sea require elaborate methods and technologies, which has been the main reason why its exploration has a comparatively short history. In the following, important key stones of deep sea exploration are listed.

- 1521: Ferdinand Magellan dropped a 700 m (2300 ft) long rope from his ship, which did not reach the ground and concluded that the sea was of infinite depth.
- 1818: The British researche Sir John Ross was the first to find that the deep sea is inhabited by life when catching jelly fish and worms in about 2000 m (6550 ft) depth with a special device.
- 1843: Nevertheless, Edward Forbes claimed that diversity of life in the deep sea is little and decreases with increasing depth. He stated that there could be no life in waters deeper than 550 m (1800 ft), the so-called Abyssus Theory.
- 1850: Near the Lofoten, Michael Sars found a rich deep sea fauna in a depth of 800 m (2600 ft) thereby refuting the Abyssus Theory.

- 1872–1876: The first systematic deep sea exploration was conducted by the Challenger Expedition on board the ship HMS *Challenger* led by Charles Wyville Thomson. This expedition revealed that the deep sea harbours a diverse, specialized biota.
- 1890–1898: First Austrian-Hungarian deep sea expedition on board the ship Pola led by Franz Steindachner in the eastern Mediterranean and the Red Sea.
- 1898–1899: First German deep sea expedition on board the ship Valdivia led by Carl Chun; found many new species from depths greater than 4000 m (13000 ft) in the southern Atlantic Ocean.
- 1930: William Beebe and Otis Barton are the first humans to reach the Deep Sea when diving in the so-called Bathysphere, made from steel. They reach a depth of 435 m (1430 ft), where they observed jelly fish and shrimp.
- 1934: The Bathysphere reached a depth of 923 m (3028 ft).
- 1948: Otis Barton set out for a new record reaching a depth of 1370 m (4495 ft).
- 1960: Jacques Piccard and Don Walsh reached the deepest point at 10,740 m (35236 ft), in their deep sea vessel Trieste, where they observed fish and other deep sea organisms.

Oceanographic Instrumentation

The sounding weight, one of the first instruments used for the sea bottom investigation, was designed as a tube on the base which forced the seabed in when it hit the bottom of the ocean. British explorer Sir James Clark Ross fully employed this instrument to reach a depth of 3,700 m (12,140 ft) in 1840.

The sounding weights used on the HMS Challenger were slightly advanced called “Baillie sounding machine”. The British researchers used wire-line soundings to investigate sea depths and collected hundreds of biological samples from all the oceans except the Arctic. Also used on the HMS Challenger were dredges and scoops, suspended on ropes, with which samples of the sediment and biological specimens of the seabed could be obtained.

A more advanced version of the sounding weight is the gravity corer. The gravity corer allows researchers to sample and study sediment layers at the bottom of oceans. The corer consists of an open-ended tube with a lead weight and a trigger mechanism that releases the corer from its suspension cable when the corer is lowered over the seabed and a small weight touches the ground. The corer falls into the seabed and penetrates it to a depth of up to 10 m (33 ft). By lifting the corer, a long, cylindrical sample is extracted in which the structure of the seabed's layers of sediment is preserved. Recovering sediment cores allows scientists to see the presence or absence of specific fossils in the mud that may indicate climate patterns at times in the past, such as during the ice ages. Samples of deeper layers can be obtained with a corer mounted in a drill. The drilling vessel JOIDES Resolution is equipped to extract cores from depths of as much as 1,500 m (4900 ft) below the ocean bottom.

Echo-sounding instruments have also been widely used to determine the depth of the sea bottom since World War II. This instrument is used primarily for determining the depth of water by means of an acoustic echo. A pulse of sound sent from the ship is reflected from the sea bottom back to the ship, the interval of time between transmission and reception being proportional to the depth of the water. By registering the time lapses between outgoing and returning signals continuously on paper tape, a continuous mapping of the seabed is obtained. Majority of the ocean floor has been mapped in this way.

In addition, high-resolution television cameras, thermometres, pressure metres, and seismographs are other notable instruments for deep-sea exploration invented by the technological advance. These instruments are either lowered to the sea bottom by long cables or directly attached to submersible buoys. Deep-sea currents can be studied by floats carrying an ultrasonic sound device so that their movements can be tracked from aboard the research vessel. Such vessels themselves are equipped with state -of-art navigational instruments, such as satellite navigation systems, and global positioning systems that keep the vessel in a live position relative to a sonar beacon on the bottom of the ocean.

Oceanographic Submersibles

Because of the high pressure, the depth to which a diver can descend without special equipment is limited. The deepest recorded made by a skin diver is 127 metres (417 ft). The deepest record made by a scuba diver is not much deeper, at 145 metres (475 ft). Revolutionary new diving suits, such as the "JIM suit," allows divers to reach depths up to approximately 600 metres (2,000 ft). Some additional suits feature thruster packs that boost a diver to different locations underwater.

To explore even deeper depths, deep-sea explorers must rely on specially constructed steel chambers to protect them. The American explorer William Beebe, also a naturalist from Columbia University in New York, was the designer of the first practical bathysphere to observe marine species at depths that could not be reached by a diver. The Bathysphere, a spherical steel vessel, was designed by Beebe and his fellow engineer Otis Barton, an engineer at Harvard University. In 1930 Beebe and Barton reached a depth of 435 m (about 1425 ft), and 923 m (3028 ft) in 1934. The potential danger was that if the cable broke, the occupants could not return to the surface. During the dive, Beebe peered out of a porthole and reported his observations by telephone to Barton who was on the surface.

In 1948, Swiss physicist Auguste Piccard tested a much deeper-diving vessel he invented called the bathyscaphe, a navigable deep-sea vessel with its gasoline-filled float and suspended chamber or gondola of spherical steel. On an experimental dive in the Cape Verde Islands, his bathyscaphe successfully withstood the pressure on it at 1,402 metres (4,600 ft), but body was severely damaged by heavy waves after the dive. In 1954, with this bathyscaphe, Piccard reached a depth of 4,000 m (13,125 ft). In 1953, his son Jacques Piccard joined in building new and improved bathyscaphe *Trieste*, which dived to 3,139 metres (10,300 ft) in field trials.

The U.S. Navy acquired Trieste in 1958 and equipped it with a new cabin to enable it to reach deep ocean trenches. In 1960, Jacques Piccard and Navy Lieutenant Donald Walsh descended in Trieste to the deepest known point on Earth - the Challenger Deep in the Mariana Trench, successfully making the deepest dive in history: 10,915 metres (35,810 ft). An increasing number of occupied submersibles are now employed around the world. The American-built DSV *Alvin* that is operated by the Woods Hole Oceanographic Institution, is a three-person submarine that can dive to about 3,600 m (12,000 ft) and is equipped with a mechanical manipulator to collect bottom samples. Alvin made its first test dive in 1964, and has performed more than 3,000 dives to average depths of 1,829 metres (6,000 ft). Alvin has also involved in a wide variety of research projects, such as one where giant tube worms were discovered on the Pacific Ocean floor near the Galápagos Islands.

Unmanned Submersibles

ROVs, or Remote Operated Vehicles, are seeing increasing use in underwater exploration. These submersibles are piloted through a cable which connects to the surface ship, and they can reach depths of up to 6,000 metres. New developments in robotics have also led to the creation of AUVs, or Autonomous Underwater Vehicles. The robotic submarines are programmed in advance, and receive no instruction from the surface. HROV combine features of both ROVs and AUV, operating independently or with a cable. Argo was employed in 1985 to locate the wreck of the *RMS Titanic*; the smaller *Jason* was also used to explore the ship wreck.

Scientific Results

In 1974 the Alvin (operated by the Woods Hole Oceanographic Institution), the French bathyscaphe Archimède, and the French diving saucer Cyane, assisted by support ships and the Glomar Challenger, explored the great Rift Valley of the Mid-Atlantic Ridge, southwest of the Azores. About 5,200 photographs of the region were taken, and samples of relatively young solidified magma were found on each side of the central fissure of the Rift Valley, giving additional proof that the seafloor spreads at this site at a rate of about 2.5 cm (about 1 in) per year. In a series of dives conducted between 1979–1980 into the Galápagos rift, off the coast of Ecuador, French, Italian, Mexican, and U.S. scientists found vents, nearly 9 m (nearly 30 ft) high and about 3.7 m (about 12 ft) across, discharging a mixture of hot water (up to 300°C/570°F) and dissolved metals in dark, smoke-like plumes. These hot springs play an important role in the formation of deposits that are enriched in copper, nickel, cadmium, chromium, and uranium.

Paleoceanography

paleoceanography, scientific study of Earth's oceanographic history involving the analysis of the ocean's sedimentary record, the history of tectonic plate motions, glacial changes, and established relationships between present sedimentation patterns and environmental factors.

Prior to the breakup of Pangea, one enormous ocean, Panthalassa, existed on Earth. Currents in this ocean would have been simple and slow, and Earth's climate was, in all likelihood, warmer than today. The Tethys seaway formed as Pangea broke into Gondwana and Laurasia. In the narrow ocean basins of the central North Atlantic, restricted ocean circulation favoured deposition of evaporites (halite, gypsum, anhydrite, and other less abundant salts). Evaporites also were deposited some 100 million years ago in the equatorial regions of the South Atlantic during the early opening of this ocean.

Sequences of organic-rich, black shales were deposited during the early phases of spreading in the North and South Atlantic. These sediments indicate anoxic conditions in the deep ocean waters. The oceans must have been well stratified into dense layers to prevent the overturning and mixing required to replace depleted oxygen. Black shales also were deposited in the older areas of the eastern Indian Ocean.

During the time interval between 200 and 65 million years ago, but especially from about 100 to 65 million years ago, microplankton abundance and diversity increased enormously in the oceans. This resulted in increased deposition of biogenic sediments in the ocean basin. During the Cretaceous Period (145.5 to 65.5 million years ago), sea level was often high, and shallow seas lapped onto the continents. This may have provided an environment favourable to the explosion in the numbers of species of foraminiferans, diatoms, and calcareous nannoplankton (single-celled, photosynthetic organisms with shells made up of calcium carbonate plates called coccoliths). Increased abundance of calcareous nannoplankton shifted the locus of carbonate sedimentation from shallow seas to the deep ocean. The end of the Cretaceous Period is marked by a sudden extinction of many life-forms on Earth, and marine organisms were no exception. Coccolithophores (calcareous nannoplankton) and planktonic foraminiferans were particularly affected, and only a few species survived. Ocean sediments were suddenly less biogenic, and clays became widespread.

After the Cretaceous Period Earth underwent a gradual cooling, especially at high latitudes. Deep-sea sedimentation changed as thermohaline bottom-water circulation became fully developed. An event of major significance was the spreading away of Australia from Antarctica beginning about 58 million years ago. This separation initiated limited circum-Antarctic circulation, which isolated Antarctica from the warmer oceans to the north, and led to cooling, which set the stage for later major glaciation.

At the boundary between the Eocene and Oligocene epochs (33.9 million years ago), Antarctic Bottom Water (AABW) began to form, resulting in greatly decreased bottom-water temperatures in both the

Pacific and Atlantic oceans. Bottom-living organisms were strongly affected, and the CCD suddenly dropped from about 3,500 metres (about 11,500 feet) to approximately 4,000 to 5,000 metres (13,000 to 16,000 feet) in the Pacific. Bottom-water temperatures were generally warm, 12 to 15 °C (54 to 59 °F), during the time preceding this event. In a study of deep-sea sediment core material from near Antarctica, New Zealand Earth scientist J.P. Kennett and American oceanographer Lowell D. Stott discovered that there was a period between roughly 50 and 35 million years ago when deep waters were very warm (20 °C [68 °F]) and salty. The origin of these ocean waters was most likely in the low latitudes and resulted from high evaporation rates there.

The modern oceans are distinguished by very cold bottom water. The gradual changes toward this condition began 10 million years after the origination of AABW. Particularly significant among these changes was the closing of the Tethys seaway as Australia and several microcontinents moved north into the Indonesian region. Also, Australia moved far enough north that circum-Antarctic surface circulation became fully established. The modern ocean circulation patterns and basin shapes were mostly in place by the beginning of the Miocene Epoch (about 23 million years ago). An exception was an ocean connection between the Pacific and Caribbean Sea in Central America that persisted until about three million years ago. Major and probably permanent ice sheets on Antarctica formed during the Miocene Epoch, and glacial sediments began to dominate the seafloor surrounding the continent shortly thereafter. Siliceous oozes also became widespread around Antarctica. Siliceous sedimentation increased in this area at the expense of siliceous sedimentation in equatorial regions. Ocean circulation became more vigourous, global climate became cooler, and sedimentation rates in the ocean basins increased. Planktonic microorganisms were segregated into latitudinal belts. Also during the Miocene Epoch rifting between Greenland and Europe had progressed to a point where a connection was established between the North Atlantic and the Norwegian Sea. This resulted in the formation of North Atlantic Deep Water, which began flowing south along the continental rise of North America at this time. Sediments redistributed and deposited by this deep current are called contourites and have been extensively studied by American geologists Bruce Heezen, Charles D. Hollister, and Brian E. Tucholke, among others.

Sudden global cooling set in near the end of the Miocene Epoch some six million years ago. The strength of ocean circulation must have increased, as evidence of increased upwelling and biological productivity is present in ocean sediments. Diatomaceous sediments were deposited in abundance around the rim of the Pacific. This cooling event is synchronous with a drop in sea level, thought to be about 40 or 50 metres (130 to 165 feet) by various authorities, and probably corresponds to the further growth of the Antarctic ice sheet. This lowered sea level, coupled with the closure of narrow seaways probably due to plate movements, isolated the Mediterranean Sea. Subsequently, the sea dried up, leaving evaporite deposits on its floor. The Swiss geologist Kenneth J. Hsü and the American oceanographer William B.F. Ryan have concluded that the Mediterranean probably dried up about 40 times as seaways opened and closed between six and five million years ago. This evaporation removed about 6 percent of the salt from the world ocean, which raised the freezing point of seawater and promoted further growth of the sea ice surrounding Antarctica.

Enormous ice sheets emerged in the Northern Hemisphere between three and two million years ago, and the succession of Quaternary glaciations began at 1.6 million years ago. The exact cause of the glacial period is unclear, but it is most likely related to the variability in solar isolation, increased mountain building, and an intensification of the Gulf Stream at three million years ago due to the closing off of the Pacific-Caribbean ocean connection in Central America. The Quaternary glaciations, of which there were probably 30 episodes, left the most dramatic record in ocean sediments of any event in the previous 200 million years. Biogenic sedimentation also increased and fluctuated with the glacial episodes. Deep-sea erosion began in many places as a result of intensified bottom-water circulation.

Coral Reef

Coral reefs are underwater structures made from calcium carbonate secreted by corals. Coral reefs are colonies of tiny living animals found in marine waters that contain few nutrients. Most coral reefs are built from stony corals, which in turn consist of polyps that cluster in groups. The polyps are like tiny sea anemones, to which they are closely related. Unlike sea anemones, coral polyps secrete hard carbonate exoskeletons which support and protect their bodies. Reefs grow best in warm, shallow, clear, sunny and agitated waters.

Often called "rainforests of the sea", coral reefs form some of the most diverse ecosystems on Earth. They occupy less than 0.1% of the world's ocean surface, about half the area of France, yet they provide a home for 25% of all marine species, including fish, mollusks, worms, crustaceans, echinoderms, sponges, tunicates and other cnidarians. Paradoxically, coral reefs flourish even though they are surrounded by ocean waters that provide few nutrients. They are most commonly found at shallow depths in tropical waters, but deep water and cold water corals also exist on smaller scales in other areas.

Coral reefs deliver ecosystem services to tourism, fisheries and shoreline protection. The annual global economic value of coral reefs has been estimated at $US375 billion. However, coral reefs are fragile ecosystems, partly because they are very sensitive to water temperature. They are under threat from climate change, ocean acidification, blast fishing, cyanide fishing for aquarium fish, overuse of reef resources, and harmful land-use practices, including urban and agricultural runoff and water pollution, which can harm reefs by encouraging excess algae growth.

Formation

Most coral reefs were formed after the last glacial period when melting ice caused the sea level to rise and flood the continental shelves. This means that most coral reefs are less than 10,000 years old. As communities established themselves on the shelves, the reefs grew upwards, pacing rising sea levels. Reefs that rose too slowly could become *drowned reefs*, covered by so much water that there was insufficient light. Coral reefs are found in the deep sea away from continental shelves, around oceanic islands and as atolls. The vast majority of these islands are volcanic in origin. The few exceptions have tectonic origins where plate movements have lifted the deep ocean floor on the surface.

In 1842 in his first monograph, *The Structure and Distribution of Coral Reefs* Charles Darwin set out his theory of the formation of atoll reefs, an idea he conceived during the voyage of the *Beagle*. He theorised uplift and subsidence of the Earth's crust under the oceans formed the atolls. Darwin's theory sets out a sequence of three stages in atoll formation. It starts with a fringing reef forming around an extinct volcanic island as the island and ocean floor subsides. As the subsidence continues, the fringing reef becomes a barrier reef, and ultimately an atoll reef.

- Darwin's theory starts with a volcanic island which becomes extinct

- As the island and ocean floor subside, coral growth builds a fringing reef, often including a shallow lagoon between the land and the main reef.
- As the subsidence continues, the fringing reef becomes a larger barrier reef further from the shore with a bigger and deeper lagoon inside.
- Ultimately, the island sinks below the sea, and the barrier reef becomes an atoll enclosing an open lagoon.

Darwin predicted that underneath each lagoon would be a bed rock base, the remains of the original volcano. Subsequent drilling proved this correct. Darwin's theory followed from his understanding that coral polyps thrive in the clean seas of the tropics where the water is agitated, but can only live within a limited depth range, starting just below low tide. Where the level of the underlying earth allows, the corals grow around the coast to form what he called fringing reefs, and can eventually grow out from the shore to become a barrier reef.

Where the bottom is rising, fringing reefs can grow around the coast, but coral raised above sea level dies and becomes white limestone. If the land subsides slowly, the fringing reefs keep pace by growing upwards on a base of older, dead coral, forming a barrier reef enclosing a lagoon between the reef and the land. A barrier reef can encircle an island, and once the island sinks below sea level a roughly circular atoll of growing coral continues to keep up with the sea level, forming a central lagoon. Barrier reefs and atolls do not usually form complete circles, but are broken in places by storms. Like sea level rise, a rapidly subsiding bottom subside can overwhelm coral growth, killing the animals and the reef.

The two main variables determining the geomorphology, or shape, of coral reefs are the nature of the underlying substrate on which they rest, and the history of the change in sea level relative to that substrate.

The approximately 20,000 year old Great Barrier Reef offers an example of how coral reefs formed on continental shelves. Sea level was then 120 m (390 ft) lower than in the 21st century. As sea level rose, the water and the corals encroached on what had been hills of the Australian coastal plain. By 13,000 years ago, sea level had risen to 60 m (200 ft) lower than at present, and many hills of the coastal plains had become continental islands. As the sea level rise continued, water topped most of the continental islands. The corals could then overgrow the hills, forming the present cays and reefs. Sea level on the Great Barrier Reef has not changed significantly in the last 6,000 years, and the age of the modern living reef structure is estimated to be between 6,000 and 8,000 years. Although the Great Barrier Reef formed along a continental shelf, and not around a volcanic island, Darwin's principles apply.

Development stopped at the barrier reef stage, since Australia is not about to submerge. It formed the world's largest barrier reef, 300–1,000 m (980–3,300 ft) from shore, stretching for 2,000 km (1,200 mi). Healthy tropical coral reefs grow horizontally from 1 to 3 cm (0.39 to 1.2 in) per year, and grow vertically anywhere from 1 to 25 cm (0.39 to 9.8 in) per year; however, they grow only at depths shallower than 150 m (490 ft) because of their need for sunlight, and cannot grow above sea level.

Types

The three principal reef types are:

- Fringing reef – this type is directly attached to a shore, or borders it with an intervening shallow channel or lagoon.
- Barrier reef – a reef separated from a mainland or island shore by a deep channel or lagoon
- Atoll reef – this more or less circular or continuous barrier reef extends all the way around a lagoon without a central island.

Other reef types or variants are:

- Patch reef – this type is an isolated, comparatively small reef outcrop, usually within a lagoon or embayment, often circular and surrounded by sand or seagrass. Patch reefs are common.
- Apron reef – a short reef resembling a fringing reef, but more sloped; extending out and downward from a point or peninsular shore
- Bank reef – a linear or semicircular shaped-outline, larger than a patch reef
- Ribbon reef – a long, narrow, possibly winding reef, usually associated with an atoll lagoon
- Table reef – an isolated reef, approaching an atoll type, but without a lagoon
- Habili – this is a reef in the Red Sea that does not reach the surface near enough to cause visible surf, although it may be a hazard to ships (from the Arabic for "unborn").

- Microatoll – certain species of corals form communities called microatolls. The vertical growth of microatolls is limited by average tidal height. By analysing growth morphologies, microatolls offer a low-resolution record of patterns of sea level change. Fossilized microatolls can also be dated using radioactive carbon dating. Such methods have been used to reconstruct Holocene sea levels.
- Cays – are small, low-elevation, sandy islands formed on the surface of coral reefs. Material eroded from the reef piles up on parts of the reef or lagoon, forming an area above sea level. Plants can stabilise cays enough to become habitable by humans. Cays occur in tropical environments throughout the Pacific, Atlantic and Indian Oceans (including the Caribbean and on the Great Barrier Reef and Belize Barrier Reef), where they provide habitable and agricultural land for hundreds of thousands of people.
- When a coral reef cannot keep up with the sinking of a volcanic island, a seamount or guyot is formed. The tops of seamounts and guyots are below the surface. Seamounts are rounded at the top and guyots are flat. The flat top of the guyot, also called a *tablemount*, is due to erosion by waves, winds, and atmospheric processes.

Zones

Coral reef ecosystems contain distinct zones that represent different kinds of habitats. Usually, three major zones are recognized: the fore reef, reef crest, and the back reef (frequently referred to as the reef lagoon).

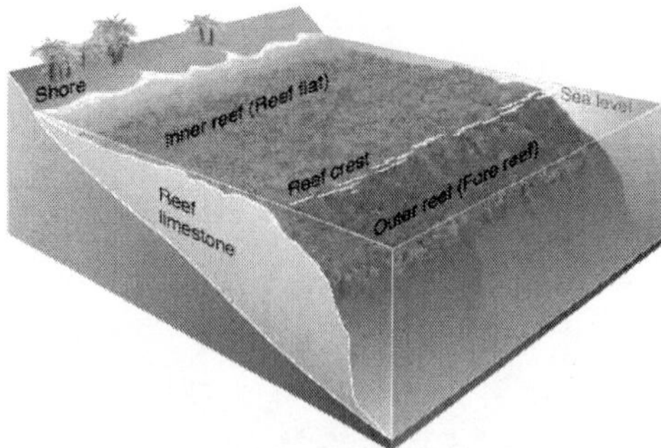

Figure: *The three major zones of a coral reef: the fore reef, reef crest, and the back reef*

All three zones are physically and ecologically interconnected. Reef life and oceanic processes create opportunities for exchange of seawater, sediments, nutrients, and marine life among one another.

Thus, they are integrated components of the coral reef ecosystem, each playing a role in the support of the reefs' abundant and diverse fish assemblages.

Most coral reefs exist in shallow waters less than 50 m deep. Some inhabit tropical continental shelves where cool, nutrient rich upwelling does not occur, such as Great Barrier Reef. Others are found in the deep ocean surrounding islands or as atolls, such as in the Maldives. The reefs surrounding islands form when islands subside into the ocean, and atolls form when an island subsides below the surface of the sea.

Alternatively, Moyle and Cech distinguish six zones, though most reefs possess only some of the zones.

When a wave enters shallow water it shoals, that is, it slows down and the wave height increases.

- The reef surface is the shallowest part of the reef. It is subject to the surge and the rise and fall of tides. When waves pass over shallow areas, they shoal, as shown in the diagram at the right. This means the water is often agitated. These are the precise condition under which corals flourish. Shallowness means there is plenty of light for photosynthesis by the symbiotic zooxanthellae, and agitated water promotes the ability of coral to feed on plankton. However, other organisms must be able to withstand the robust conditions to flourish in this zone.
- The off-reef floor is the shallow sea floor surrounding a reef. This zone occurs by reefs on continental shelves. Reefs around tropical islands and atolls drop abruptly to great depths, and do not have a floor. Usually sandy, the floor often supports seagrass meadows which are important foraging areas for reef fish.
- The reef drop-off is, for its first 50 m, habitat for many reef fish who find shelter on the cliff face and plankton in the water nearby. The drop-off zone applies mainly to the reefs surrounding oceanic islands and atolls.
- The reef face is the zone above the reef floor or the reef drop-off. "It is usually the richest habitat. Its complex growths of

coral and calcareous algae provide cracks and crevices for protection, and the abundant invertebrates and epiphytic algae provide an ample source of food."

- The reef flat is the sandy-bottomed flat can be behind the main reef, containing chunks of coral. "The reef flat may be a protective area bordering a lagoon, or it may be a flat, rocky area between the reef and the shore. In the former case, the number of fish species living in the area often is the highest of any reef zone."
- The reef lagoon – "many coral reefs completely enclose an area, thereby creating a quiet-water lagoon that usually contains small patches of reef."

However, the "topography of coral reefs is constantly changing. Each reef is made up of irregular patches of algae, sessile invertebrates, and bare rock and sand.

The size, shape and relative abundance of these patches changes from year to year in response to the various factors that favour one type of patch over another. Growing coral, for example, produces constant change in the fine structure of reefs. On a larger scale, tropical storms may knock out large sections of reef and cause boulders on sandy areas to move."

Locations

Coral reefs are estimated to cover 284,300 km^2 (109,800 sq mi), just under 0.1% of the oceans' surface area. The Indo-Pacific region (including the Red Sea, Indian Ocean, Southeast Asia and the Pacific) account for 91.9% of this total. Southeast Asia accounts for 32.3% of that figure, while the Pacific including Australia accounts for 40.8%. Atlantic and Caribbean coral reefs account for 7.6%.

Although corals exist both in temperate and tropical waters, shallow-water reefs form only in a zone extending from 30° N to 30° S of the equator. Tropical corals do not grow at depths of over 50 metres (160 ft). The optimum temperature for most coral reefs is 26–27 °C (79–81 °F), and few reefs exist in waters below 18 °C (64 °F). However, reefs in the Persian Gulf have adapted to temperatures of 13 °C (55 °F) in winter and 38 °C (100 °F) in summer.

Deep water coral can exist at greater depths and colder temperatures at much higher latitudes, as far north as Norway. Although deep water corals can form reefs, very little is known about them. Coral reefs are rare along the American and African west coasts. This is due primarily to upwelling and strong cold coastal currents that reduce water temperatures in these areas (respectively the Peru, Benguela and Canary streams). Corals are seldom found along the coastline of South Asia from the eastern tip of India (Madras) to the Bangladesh and Myanmar borders. They are also rare along the coast around northeastern South America and Bangladesh due to the freshwater release from the Amazon and Ganges Rivers, respectively.

- The Great Barrier Reef—largest, comprising over 2,900 individual reefs and 900 islands stretching for over 2,600 kilometres (1,600 mi) off Queensland, Australia
- The Mesoamerican Barrier Reef System—second largest, stretching 1,000 kilometres (620 mi) from Isla Contoy at the tip of the Yucatán Peninsula down to the Bay Islands of Honduras
- The New Caledonia Barrier Reef—second longest double barrier reef, covering 1,500 kilometres (930 mi)
- The Andros, Bahamas Barrier Reef—third largest, following the east coast of Andros Island, Bahamas, between Andros and Nassau
- The Red Sea—includes 6000-year-old fringing reefs located around a 2,000 km (1,240 mi) coastline
- Pulley Ridge—deepest photosynthetic coral reef, Florida
- Numerous reefs scattered over the Maldives
- The Raja Ampat Islands in Indonesia's West Papua province offer the highest known marine diversity.

Biology

Live coral are small animals embedded in calcium carbonate shells. It is a mistake to think of coral as plants or rocks. Coral heads consist of accumulations of individual animals called polyps, arranged in diverse shapes. Polyps are usually tiny, but they can range in size from a pinhead to 12 inches (30 cm) across.

Reef-building or hermatypic corals live only in the photic zone (above 50 m), the depth to which sufficient sunlight penetrates the water, allowing photosynthesis to occur. Coral polyps do not themselves photosynthesize, but have a symbiotic relationship with zooxanthellae;

these organisms live within the tissues of polyps and provide organic nutrients that nourish the polyp. Because of this relationship, coral reefs grow much faster in clear water, which admits more sunlight. Without their symbionts, coral growth would be too slow for the corals to form significant reef structures. Corals get up to 90% of their nutrients from their symbionts.

Reefs grow as polyps and other organisms deposit calcium carbonate, the basis of coral, as a skeletal structure beneath and around themselves, pushing the coral head's top upwards and outwards. Waves, grazing fish (such as parrotfish), sea urchins, sponges, and other forces and organisms act as bioeroders, breaking down coral skeletons into fragments that settle into spaces in the reef structure or form sandy bottoms in associated reef lagoons. Many other organisms living in the reef community contribute skeletal calcium carbonate in the same manner. Coralline algae are important contributors to reef structure in those parts of the reef subjected to the greatest forces by waves (such as the reef front facing the open ocean). These algae strengthen the reef structure by depositing limestone in sheets over the reef surface.

The colonies of the one thousand coral species assume a characteristic shape such as wrinkled brains, cabbages, table tops, antlers, wire strands and pillars.

Corals reproduce both sexually and asexually. An individual polyp uses both reproductive modes within its lifetime. Corals reproduce sexually by either internal or external fertilization. The reproductive cells are found on the mesentery membranes that radiate inward from the layer of tissue that lines the stomach cavity. Some mature adult corals are hermaphroditic; others are exclusively male or female. A few species change sex as they grow.

Internally fertilized eggs develop in the polyp for a period ranging from days to weeks. Subsequent development produces a tiny larva, known as a planula. Externally fertilized eggs develop during synchronised spawning. Polyps release eggs and sperm into the water en masse, simultaneously. Eggs disperse over a large area. The timing of spawning depends on time of year, water temperature, and tidal and lunar cycles. Spawning is most successful when there is little variation between high and low tide. The less water movement, the better the chance for fertilization. Ideal timing occurs in the spring. Release of eggs or planula usually occurs at night, and is sometimes in phase with the lunar cycle (three to six days after a full moon). The period from release to settlement lasts only a few days, but some planulae can survive afloat for several weeks. They are vulnerable to predation and environmental conditions. The lucky few planulae which successfully attach to substrate next confront competition for food and space.

Darwin's Paradox

Coral... seems to proliferate when ocean waters are warm, poor, clear and agitated, a fact which Darwin had already noted when he passed through Tahiti in 1842.

This constitutes a fundamental paradox, shown quantitatively by the apparent impossibility of balancing input and output of the nutritive elements which control the coral polyp metabolism.

Recent oceanographic research has brought to light the reality of this paradox by confirming that the oligotrophy of the ocean euphotic zone persists right up to the swell-battered reef crest. When you approach the reef edges and atolls from the quasidesert of the open sea, the near absence of living matter suddenly becomes a plethora of life, without transition. So why is there something rather than nothing, and more precisely, where do the necessary nutrients for the functioning of this extraordinary coral reef machine come from ? — Francis Rougerie

During his voyage on the *Beagle*, Darwin described tropical coral reefs as oases in the desert of the ocean. He reflected on the paradox that tropical coral reefs, which are among the richest and most diverse ecosystems on earth, flourish surrounded by tropical ocean waters that provide hardly any nutrients.

Coral reefs cover less than 0.1% of the surface of the world's ocean, yet they support over one-quarter of all marine species. This diversity results in complex food webs, with large predator fish eating smaller forage fish that eat yet smaller zooplankton and so on. However, all food webs eventually depend on plants, which are the primary producers. Coral reefs' primary productivity is very high, typically producing 5–10 g $\cdot cm^{-2} \cdot day^{-1}$ biomass.

One reason for the unusual clarity of tropical waters is they are deficient in nutrients and drifting plankton. Further, the sun shines year round in the tropics, warming the surface layer, making it less dense than subsurface layers. The warmer water is separated from deeper, cooler water by a stable thermocline, where the temperature makes a rapid change. This keeps the warm surface waters floating above the cooler deeper waters. In most parts of the ocean, there is little exchange between these layers. Organisms that die in aquatic environments generally sink to the bottom, where they decompose, which releases nutrients in the form of nitrogen (N), phosphorus (P) and potassium (K). These nutrients are necessary for plant growth, but in the tropics, they do not directly return to the surface.

Plants form the base of the food chain, and need sunlight and nutrients to grow. In the ocean, these plants are mainly microscopic phytoplankton which drift in the water column. They need sunlight for photosynthesis, which powers carbon fixation, so they are found only relatively near the surface. But they also need nutrients. Phytoplankton rapidly use nutrients in the surface waters, and in the tropics, these nutrients are not usually replaced because of the thermocline.

Coral Polyps

Around coral reefs, lagoons fill in with material eroded from the reef and the island. They become havens for marine life, providing protection from waves and storms.

Most importantly, reefs recycle nutrients, which happens much less in the open ocean. In coral reefs and lagoons, producers include phytoplankton, as well as seaweed and coralline algae, especially small types called turf algae, which pass nutrients to corals. The phytoplankton are eaten by fish and crustaceans, who also pass nutrients along the food web. Recycling ensures fewer nutrients are needed overall to support the community.

Coral reefs support many symbiotic relationships. In particular, zooxanthellae provide energy to coral in the form of glucose, glycerol, and amino acids. Zooxanthellae can provide up to 90% of a coral's energy requirements. In return, as an example of mutualism, the corals shelter the zooxanthellae, averaging one million for every cubic centimetre of coral, and provide a constant supply of the carbon dioxide they need for photosynthesis.

Corals also absorb nutrients, including inorganic nitrogen and phosphorus, directly from water. Many corals extend their tentacles at night to catch zooplankton that brush them when the water is agitated. Zooplankton provide the polyp with nitrogen, and the polyp shares some of the nitrogen with the zooxanthellae, which also require this element. The varying pigments in different species of zooxanthellae give them an overall brown or golden-brown appearance, and give brown corals their colours. Other pigments such as reds, blues, greens, etc. come from coloured proteins made by the coral animals. Coral which loses a large fraction of its zooxanthellae becomes white (or sometimes pastel shades in corals that are richly pigmented with their own colourful proteins) and is said to be bleached, a condition which, unless corrected, can kill the coral.

Sponges are another key to explaining Darwin's paradox. They live in crevices in the coral reefs. They are efficient filter feeders, and in the Red Sea they consume about 60% of the phytoplankton that drifts by. The sponges eventually excrete nutrients in a form the corals can use.

The roughness of coral surfaces is the key to coral survival in agitated waters. Normally, a boundary layer of still water surrounds a submerged object, which acts as a barrier. Waves breaking on the extremely rough edges of corals disrupt the boundary layer, allowing the corals access to passing nutrients. Turbulent water thereby promotes reef growth and branching. Without the nutritional gains brought by rough coral surfaces, even the most effective recycling would leave corals wanting in nutrients.

Cyanobacteria provide soluble nitrates for the reef via nitrogen fixation.

Coral reefs also often depend on surrounding habitats, such as seagrass meadows and mangrove forests, for nutrients. Seagrass and mangroves supply dead plants and animals which are rich in nitrogen and also serve to feed fish and animals from the reef by supplying wood and vegetation. Reefs, in turn, protect mangroves and seagrass from waves and produce sediment in which the mangroves and seagrass can root.

Biodiversity

Coral reefs form some of the world's most productive ecosystems, providing complex and varied marine habitats that support a wide range of other organisms. Fringing reefs just below low tide level also have a mutually beneficial relationship with mangrove forests at high tide level and sea grass meadows in between: the reefs protect the mangroves and seagrass from strong currents and waves that would damage them or erode the sediments in which they are rooted, while the mangroves and sea grass protect the coral from large influxes of silt, fresh water and pollutants. This additional level of variety in the environment is beneficial to many types of coral reef animals, which, for example, may feed in the sea grass and use the reefs for protection or breeding.

Reefs are home to a large variety of organisms, including fish, seabirds, sponges, cnidarians (which includes some types of corals and jellyfish), worms, crustaceans (including shrimp, cleaner shrimp, spiny lobsters and crabs), mollusks (including cephalopods), echinoderms (including starfish, sea urchins and sea cucumbers), sea squirts, sea turtles and sea snakes. Aside from humans, mammals are rare on coral reefs, with visiting cetaceans such as dolphins being the main exception. A few of these varied species feed directly on corals, while others graze on algae on the reef. Reef biomass is positively related to species diversity.

Technologies and Capabilities for Ocean Research and Exploration

NOAA's various missions in coastal and ocean waters depend on technology. It is only in the past half century that technology has advanced to the point that we can examine the oceans in systematic, scientific and noninvasive ways. The changing and difficult study of the ocean realm requires new intellectual approaches and a national investment in a new mode of conducting marine investigations. With the exception of advanced synoptic remote sensing technologies over the last several decades, surface-ship expeditions have dominated ocean science since World War II. New approaches, such as advanced sea floor observatories and human-occupied habitats greatly enhance traditional capabilities by providing invaluable long-term monitoring and continuity of observations.

Why is Ocean Technology Important?

Technology should be viewed as a means to support research and exploration on and under the ocean and to support NOAA's broad responsibilities in understanding and predicting changes in our environment. New technologies are also utilised to conserve and manage our coastal and marine resources to meet the Nation's needs environmentally and economically. The numerous technologies that make this possible include the ships, submersibles, diving technologies and observation tools that transport us across oceans and into their depths allowing us to examine, record, and conduct research on the multitude of mysteries. NOAA Research, along with its extramural partners, owns or provides access to a wide range of assets that can conduct operations from the coastal oceans and large lakes to depths of 6,500 metres (21,300 feet), accomplishing a broad spectrum of research tasks. These include both occupied and unoccupied vehicles as well as platforms and observatories such as the Aquarius laboratory and the LEO-15 underwater observatory. In addition we operate numerous small ROVs and a light-work ROV, which has substantial payload and sampling capability to depths to 1,000 metres. We are in the process of developing several autonomous undersea vehicle (AUV) designs that will serve NOAA's research needs. In addition, another university partner, the Hawaii Undersea Research Laboratory, operates two occupied submersibles, Pisces IV& V, capable of 2,000 metre depths from a 220 foot support ship. We also have an agreement with the US Navy for access to the two-scientist nuclear powered submarine, NR-1. An interagency agreement with Woods Hole Oceanographic Institution provides scientists the Alvin submersible (4,500 metre depth) and Jason ROV (6,000 metres). We have also leased deep water systems from Harbor Branch Oceanographic Institution and Delta Oceanographics.

Exploration and research on ecosystems requires detailed study and observation. Working in the oceans presents unique challenges since the ocean is one of the most complex and harsh environments on earth and accessing it requires unique technologies and capabilities. Instrumentation and certain tools are needed to adequately collect data and samples from the coast to the deepest ocean depths. These include sondes, CTDs, and submersible samplers and collectors for specific needs such as geothermal fluids. Another important requirement is to scientifically document the undersea environment using still and video imagery. Specialized water and pressure resistant equipment must be developed and utilised. Underwater photography is constrained by factors unique to the aquatic environment such as very limited visibility, lighting, and water quality. Newly developed digital cameras allow for several orders of magnitude more storage and better control of the imagery being documented.

Priority Areas for Future Technology Developments

Recent technical advancements in low-power miniaturized components enable development of ocean floor stations that feature a wide variety of in situ sampling tools and sensors. Autonomous underwater vehicles (AUVs) can use sea floor observatories as a home base to power-up and download acquired data, thereby extending the geographic range of these stations. Real-time data and imagery are routinely transmitted to land-based laboratories and the internet via cables, radios, and satellite. Scientists can control in situ experiments and equipment from their land-based laboratories through sea floor observatories and habitats. The following are examples of the next generation in the development of technology for better understanding the oceans. Laser line scanning for habitat assessment provides efficiency and spatial coverage of a remote survey system and image resolution approaching visual observations. This technology allows millemetre scale resolution at two to five times the range of conventional imaging systems. Preliminary results indicate that this new capability has the potential to improve the efficiency and accuracy of fish habitat assessment. NOAA Research has partnered with universities in numerous marine biotechnology ventures. Several AUVs for science are being developed to support various NOAA missions. Preliminary requirements for one major AUV design include 1,500 metre depth capability for two days for habitat characterisation, seafloor mapping, and observatory support. In addition, the AUV will have multiple sensors and precise navigation and tracking. These technology developments will provide NOAA and its extramural partners the capability to conduct research and exploration in order to better understand our marine environment.

Monitoring Oil Slicks From Space to Aid in Offshore Petroleum Exploration

Offshore hydrocarbon reservoirs can experience oil and gas seepage through the seafloor and water column, resulting in telltale slicks on the sea surface. Detecting and identifying these intermittent and often remote slicks can contribute to directing exploration resources in both producing and frontier basins, as active oil seepage provides information about source rock type and maturity, location of source kitchens and migration pathways. These form part of a petroleum prospectivity analysis of a basin, which Geoscience Australia presents to industry to reduce its exploration risk. This type of precompetitive information also forms part of the offshore acreage releases.

Remote sensing, the acquisition of image data from sensors mounted on airborne or spaceborne platforms, is seen as a promising tool for detecting, mapping and identifying these seepage-derived slicks. Simply put, it offers a cost-effective means of scanning extensive and/or remote regions on an ongoing basis.

Challenges of Remote Sensing

The bulk of remote sensing of marine hydrocarbon slicks to date has focused on detecting and mapping relatively thick oil pollution and spills, often with a known extent, date and/or location. Utilising remote sensing for petroleum exploration purposes poses significant additional challenges: Natural, seepage-derived oil slicks are typically considerably thinner on the sea surface, and their timing, extent, location and chemical properties are often unknown.

One type of remote sensing, synthetic aperture radar (SAR), has been shown to effectively detect oil slicks on the sea surface. Under suitable environmental conditions, a sea-surface slick has a dampening effect on capillary waves, thereby reducing the amplitude of the wave-dependant backscatter signal from the sea surface. This is visible as a dark area (negative contrast) in SAR imagery.

In principle, SAR technology is ideally suited to scan for remote and transient natural oil slicks, as it covers large geographic areas (one scene can have a footprint of several hundred kilometres across the sea surface) and can image through clouds and at night. However, the volume of data collected when scanning an area as large as the Australian Marine Jurisdiction (AMJ) prohibits manual, operator-based inspection of the imagery. This problem is further compounded where multitemporal datasets are involved. In addition, new data are continuously being acquired.

A second important limitation of SAR data is that it only contains information in one spectral channel. SAR imagery can therefore be thought of as monochromatic (black and white). Hence, other phenomena that cause dampening of capillary waves, such as biological films, hydrodynamic effects, bathymetric features and even weather-induced artifacts, lead to a range of false-positive interpretations of oil slicks in SAR imagery.

Optical remote sensors, on the other hand, typically collect data in three or more spectral channels. As an example, DigitalGlobe's (Longmont, Colorado) spaceborne QuickBird sensor is referred to as multispectral, and it acquires information in four channels, the first three of which are broadly equivalent to the blue, green and red wavelengths of human vision (the fourth channel is a near-infrared channel). Hyperspectral sensors, such as the airborne HyMap sensor manufactured by Integrated Spectronics (Castle Hill, Australia), collect information in hundreds of spectral channels. These additional dimensions of wavelength-dependent information in optical remote sensing data theoretically allow for discrimination of false positives such as biological films, and may even allow for differentiation between oil types. Admittedly, this remains to be proven.

However, there is a trade-off with using optical remote sensing. Specifically, the spatial coverage of a typical scene is limited (e.g., one QuickBird scene covers approximately 15 by 15 kilometres of the sea surface), with costs ranging from thousands to tens of thousands of dollars per scene. In addition, optical remote sensing requires sunlight and cannot image through clouds. The latter constraint translates to potentially requiring several scene acquisitions over the same area in order to have one useful image.

In other words, it is too costly to use optical remote sensing data when attempting to monitor the AMJ for intermittent, seepage-derived slicks, particularly on an ongoing basis.

The Solution

One solution is to try and harness the advantages of these complementary remote sensing data types. To this end, Geoscience Australia is developing a two-pronged remote sensing-based approach to study seepage-derived slicks in the AMJ: first, building a semiautomated processing and classification system in order to scan large numbers of SAR scenes for potential natural slick targets, and second, investigating the potential of optical remote sensing as a diagnostic tool for further, targeted study in identified areas of interest.

SAMSSARA

The Semi-Automated Marine Slicks SAR Analysis (SAMSSARA) system, which was developed using Definiens (Munich, Germany) software, combines object-oriented data analysis and classification tools with batch-processing and thorough reporting schematics in order to provide a first screening of large SAR datasets.

An identified object with contiguous, lower values (i.e., negative contrast as defined by local statistics) is considered a potential slick candidate object. The shape of each potential object is then examined using a set of criteria that include length versus width and perimetre versus area. A natural slick typically has an elongated, thin form, with one or more twists (due to surface currents).

Identified objects are merged using a set criteria that include the object's orientation in relation to neighbouring objects and the extent and orientation of shared perimetres. Merged objects are then reclassified based on the contrast and shape rules previously described, and the process is repeated iteratively until every subset in the scene has been classified. The classification scheme currently uses three categories of potential slick targets: dark feature, moderately slick-shaped and slick-shaped, with the latter the most likely slick target. SAMSSARA retains the number of classified targets in each category for each scene and outputs these project statistics, together with thumbnail imagery and geocoordinates, in both spreadsheet and HTML format.

Finally, SAMSSARA outputs georeferenced maps (shapefiles) of all classified potential targets. This allows for spatial and contextual analysis of an identified potential slick feature, together with ancillary information, such as geological data.

In one example, SAMSSARA's output from the screening of 238 SAR scenes of the southwestern AMJ covered a sea surface area of approximately 1.35 million square kilometres. While the manual and subjective analysis of one SAR scene can take a trained operator approximately 30 minutes, the 238 scenes were objectively analysed by SAMSSARA in approximately six hours, producing an array of easily interpretable outputs.

Slick-shaped features of interest in the SAMSSARA output can be investigated in the context of ancillary data and revisited with additional SAR data from a different date. The benefit of using such multitemporal analysis is that the reoccurence of a very slick-shaped feature in both space and time rules out several false positive interpretations and supports directing additional resources, such as optical remote sensing data, to investigating this particular sea surface area.

Optical Remote Sensing

Optical remote sensing has been shown to be able to identify certain sea surface slicks through an ability to collect imagery data over several wavelength channels. One example is algal blooms, where a combination of spectral channels can detect the biological signature. Errors of commission in SAR data can therefore be reduced by eliminating these types of false positives using the additional information in optical data. The applicability of optical remote sensing data to positive identification of seepage-derived slicks is less straightforward. The absorption of light by oil is dependent on both wavelength and oil thickness; there is an increase in light absorption with increasing oil thickness, as well as an increase in absorption toward the blue end of the visible spectrum.

Petroleum is composed of complex mixtures of hydrocarbons, and the chemistry of individual oils can vary substantially. This variation in chemistry affects a variety of bulk properties, such as density and viscosity, and has a direct effect on the optical properties of oils. Thus, the chemistry of an oil can influence both the wavelength and thickness-dependent absorptive properties. Important-ly, it is these absorptive properties that offer the oil slick diagnostic potential in optical data.

This diagnostic potential was investigated through a series of small-scale laboratory experiments. Samples of oils known to occur in the AMJ were sourced from the Geoscience Australia archive, and the conditions under which a remote sensor would collect data in the environment were simulated. Known amounts of each oil type were added to a layer of water, and a hyperspectral radiometre was used to measure the resulting change in reflect-ance as a function of oil thickness. These changes in reflectance were then normalised to the response functions of remote sensing platforms likely to be used for oil slick detection.

The results confirmed that remote sensing-based oil slick detection is both thickness and oil type-dependent. What is more noteworthy is that the work demonstrated an objective method for quantifying the sensitivity of a given sensor to a given oil type at a given oil thickness. This has numerous applications. For example, if a basin is known to contain a certain oil type, the sensitivity analysis will reveal if optical remote sensing is applicable. Indeed, one of the oil types studied was shown to be undetectable up to a thickness exceeding those of typical natural slicks, while another oil type was shown to be detectable by several commercially available sensors at thicknesses of only 10 micrometres. This type of information is relevant to a prospective purchaser of optical remote sensing data, particularly given the conflicting information from the literature and data suppliers with regards to this application of optical imagery.

Building on the findings of these initial laboratory simulations, a more robust and easy-to-implement method has been developed to obtain the optical properties of a given oil type, making this type of sensitivity analysis more accessible to a range of users, provided that they have access to samples of the oil type in question.

This sensitivity analysis method gives an indication of the applicability of a given sensor to a given oil type, if the approximate layer thickness is known. In addition, this type of information can be used to start identifying oil types and/or estimating thicknesses. If an area is known to contain a certain oil type, and optical remote sensing data is able to detect a slick, assumptions can be made as to the minimum thickness of the layer. As an alternative example, the detection of a slick can be used to rule out the presence of oil types that have been assessed to be undetectable.

Future development of SAMSSARA will focus on furthering the multitemporal analysis component, since a reoccurence of a sea surface slick is a strong indicator of a seepage origin. The anticipation is that results will continue to be added to a database of oil type versus sensor sensitivity. This will provide a reference for the applicability of a given optical remote sensing platform for detecting a given type of marine oil slick, effectively directing future exploration resource deployment.

Rapidly Deployable SeaSonde for Modelling Oil Spill Response

Recent spill events such as the 2010 *Deepwater Horizon* accident in the Gulf of Mexico and the 2007 MV *COSCO Busan* collision in San Francisco Bay have highlighted the benefits and need for surface current mapping for response operations.

CODAR Ocean Sensors' SeaSonde high-frequency radar networks, which provided maps of ocean currents during these events, consisted of permanently installed stations at sites in developed areas that were easily accessible by road with access to grid power and high-speed data links. Most of the SeaSondes operating around the world are located at such sites, and those that are not generally require weeks or months of planning alternative power and communications solutions. Much of the world's coastlines are at risk from the production and transport of oil, but high-frequency radar current maps only cover a fraction of them. Resolving this problem requires a mobile rapid-response capability.The concept of mobile high-frequency radar systems is not a new one. In fact, it was one of the driving concepts behind the development of the compact Coastal Ocean Dynamics Radar system at NOAA in the 1970s and early 1980s. Others have since created their own customized mobile solutions using SeaSondes integrated into automobile trailers at both Texas A&M University and NOAA's Centre for Operational Oceanographic Products and Services. These solutions, while innovative in their ability to quickly deploy equipment once on site and for their utilisation of wireless communications and self-contained power sources, were still limited to coastal sites accessible by roads. Areas such as the Louisiana coast, the Mississippi Delta and along most of the coast of Norway, require a solution that can utilise a number of modes of transportation, including final positioning by helicopter.

In 2009, on behalf of the companies operating on the Norway's continental shelf, the Norwegian Clean Seas Association for Operating Companies and the Norwegian Coastal Administration launched Oil Spill Response 2010, a multiyear development program for oil spill response technology.

The goal of this program was to achieve significant improvements in providing continuous and effective oil spill response offshore and in coastal and shoreline areas during various weather, daylight and climatic conditions. In response to this call, CODARNOR AS, along with development partners CODAR Ocean Sensors, QUALITAS Remos (Madrid, Spain) and the Norwegian Meteorological Institute, developed a rapid-response SeaSonde for the rugged and remote Norwegian coastline.

SeaSonde Objectives

The objectives of this project were to develop a mobile SeaSonde high-frequency radar unit that can be rapidly deployed to the coast of Norway to aid in effective and efficient oil spill response.

The project aimed to accomplish this through developing a data service that provides high-quality SeaSonde-derived 2D current fields to the Norwegian Meteorological Institute in near real time. This would be used for spill drift model input and operations planning while also demonstrating that these current fields can improve operational oil spill drift model results.

Adapting the commercial SeaSonde to a rapid-response system required outfitting it in a protective housing that is easily transportable by a number of methods, including helicopter.

The mobile unit's shell was also required to be large enough to contain all equipment and tools, provide enough room for one person to sit comfortably, and be lightweight and compact.

Hardware Specifications

With this in mind, a shell was manufactured using a half-inch PVC foam core that was fibreglass-reinforced and polyester-laminated inside and out. The shell dimensions were 1.2 metres long by 1.2 metres wide and 1.5 metres high, mounted on skids that span a 1.2-by-two-metre area. For operating frequencies of 11 megahertz or higher, the unit uses a single transmit-receive antenna, which measures six to eight metres in total height with a four-metre-tall mast and whip antenna for the rest. For compactness during transport, the whip is quickly removable, and the mast was redesigned to be a two-piece hinged assembly that was mounted on the side of the shell. Once the rapid deployment unit is delivered, a single person can completely assemble and erect the antenna in about 10 minutes. Power was supplied via redundant, off-the-shelf, two-kilowatt Honda generators with uninterruptible power supply backup and power conditioning. The prototype unit required 400 watts of continuous power, but a new transmit module design has since brought this number down to about 200 watts. Communications for data transfer were provided by the Ice.net CDMA 450-megahertz cellular service available along the Norwegian coast, and a redundant satellite communications system will be available in production models.

Rapid Data Products

While designing hardware for rapid-response deployments is important, perhaps more important is devising a plan for how to provide data products to operators and responders. Spatial gaps or shadows in surface-current coverage due to coastline geometry, radio interference, nearfield antenna distortions, etc., can often be present in real-time, high-frequency current maps. Moreover, when using a conventional local combination of radial data to obtain surface current data, it is not possible to have currents close to the coast where radial vectors from the two stations are almost parallel due to purely geometrical considerations. To overcome these difficulties, project participants have implemented and tested, in a real-time configuration, a robust and well-established method based on open-boundary modal analysis (OMA) to fill spatial gaps and expand the data to the coast.

OMA relies on the decomposition of the total velocity field as a sum of divergence-free and irrotational modes as well as modes describing flow through the open boundary. The radial data from the high-frequency radar stations are then fitted to an optimal linear combination of these modes through the minimisation of a cost function. The modes depend exclusively on the domain selected, and the definition of the closed and open boundaries are computed only once and do not change in time. In addition to filling gaps and the baseline region, radial data from a single site can be used to reproduce an entire total vector field when data from another site are unavailable. Updated 2D current vector maps can be available with temporal resolutions of 10 to 60 minutes and delays due to data transfer and OMA processing of 10 to 20 minutes. On their own, quality-controlled 2D vector maps available in near real time provide vessel pilots information on the currents in the area they are working. A real added value, however, would come from using the data to improve spill predictions and, therefore, being able to plan positioning and utilisation with much more efficacy. This is specifically one of the goals of this effort.

Surface Carbon Flux

For the most part, these measurements remain within the research community. But the technology exists to use VOS and drifters to collect pCO_2 in situ measurements, and satellite ocean colour provides effective proxy data for pCO_2.

Sampling strategy and benchmark accuracies: Seek pCO_2 and total CO_2 measurements with an accuracy of ρ2-3 æatm and ρ2 æmol respectively. In situ sampling is not expected to reach threshold rates, so simply aim for enhanced VOS, mooring and drifter measurements, piggy-backing wherever possible on existing operational systems. Ancillary SST and atmospheric data are important. Aim for continuing global satellite ocean colour measurements, at 25-100 km resolution and daily coverage, with 2-10% accuracy. Development and validation of satisfactory remote sensing algorithms is important.

Time-series stations are playing a key role in research and the Ocean Climate Time-Series Workshop (Baltimore, MD, USA, March 1997) co-sponsored by GOOS, GCOS, WCRP and JGOFS (GOOS Report No. 33, GCOS Report No. 41) saw an important role in the future for such Time- Series.

Comments

Some non-biological applications (e.g. tropical ocean modelling) are using ocean colour to estimate opacity. Independently of any non-physical applications, this suggests that there is a good case for adding ocean colour to the list of needed remote sensing techniques.

Upper Ocean Temperature

In the past, upper ocean thermal networks have largely been the province of research. Making significant parts of these networks operational is one of the key themes of OOPC and remains a high-priority issue.

Characteristics desired of the processed signal:

- General large scale requirement is for 2-500 km scale bimonthly global maps of the heat content and the first few vertical modes of variability; and monthly climatologies on 1° resolution. An accuracy of ~ 0.5°C is useful.
- For ENSO forecasts: 1° latitude and 5° longitude resolution every 10 days and over 500m vertically (mixed layer depth (MLD) and ~5 vertical modes) to an accuracy 0.2-0.5°C.
- For mesoscale applications: 25-50 km resolution every 2 days over 500 m with an accuracy of around 0.5°C.
- For climate trend, better than 0.1 C/year accuracy.

Sampling strategy and benchmark accuracies:

- Maintain TOGA/WOCE broad-scale VOS sampling (1 XBT per months with 1.5° latitude and 5° longitude resolution). Priority to lines with established records, of good quality, and in regions of scientific significance (e.g., tropics, particularly outside the domain of TAO, and the TRANSPAC region).
- Maintain TOGA Pacific network, in particular TAO (OOSDP did not specify part or all of the present array, but did suggest "close to" 1994 levels). Around 4 samples every 5 days per 2° x 15° bin, with 10-15 m vertical resolution is deemed satisfactory.
- Enhanced coverage in the equatorial regions in the vicinity of sharp gradients (e.g. Kuroshio): O(18) sections per year, with 50-100 km resolution.
- Boost routine sampling of the polar regions (at broadcast mode levels)
- Use of profiling floats to implement a truly global observing system. This is a technology that is developing rapidly and real-time data are now available; sampling strategies have yet to be defined for "operational" use but a float profile per 2-300 km square every 10 days might be a feasible target. Argo,

developed under the auspices of GODAE, will become the mechanism for developing a strategy for deploying ~ 3000 floats globally for GODAE in the period 2003-5. As such it will serve as a pilot project for the longer term use of profiling floats in the GOOS/GCOS OOS.

Other Sources of Information

Clearly altimetry offers complementary data. For the tropics, it is feasible a good model plus SST and wind-forcing may be able to forecast subsurface temperature structure with useful skill. However, at the present time, there is no reason to lessen the requirements outlined above. Several groups are using empirical relationships plus assumptions about the T/S relationship to infer subsurface structure from altimetry (variously known as synthetic or pseudo XBTs). Acoustic thermometry has good potential, particularly for long-term change and in regional modelling. It seems highly unlikely that an in situ solution will be found for the mesoscale applications. Rather, it is likely a mix of moorings, XBTs and profiling floats may be used to pin-down the global, large-scale thermal structure, and a mix of altimetry, SST and colour used to specify the horizontal structure of the mesoscale field.

Trends

Profiling floats, and in particular the Argo initiative, are arousing a great deal of interest and seem to offer the one real chance for global temperature sampling (VOS are limited in terms of geographic coverage, and moorings are better suited to tropical and boundary regions). A program called PIRATA is testing TAO-like moorings in the tropical Atlantic, and the Japanese TRITON program is testing moorings for mid-latitude climate studies, and for Indian Ocean studies.

5

Marine Ecosystem

Observational Evidence

We adopt the local stratigraphic nomenclature established by Terlemez et al. (1997). The oldest rock exposed along the Gölbasi-Türkoglu Fault is the Hatay Ophiolite, which was obducted onto the northern margin of the Arabian Platform during the Maastrichtian (c. 70 Ma), having apparently formed around 90 Ma at a spreading centre within the adjacent Southern Neotethys Ocean. Ophiolite obduction was followed unconformably by deposition of the Besni Formation, consisting of multi-coloured conglomerate and sandstone, passing upwards into Middle or Upper Maastrichtian limestone. This is followed by the uppermost Maastrichtian to Palaeocene Germav Formation, consisting of marl and clayey limestone, overlain by the Eocene to Upper Miocene Midyat Group carbonates. Along the Gölbasi-Türkoglu Fault, this is typically represented by the Lutetian Hoya Formation, a particularly pure, well-lithified limestone that resists erosion and forms much of the relief of the region.

Figure: *Marine Ecosystems*

Mid-Miocene uplift and sea-level fluctuation (e.g. Karig & Kozlu 1990; Arger et al. 2000) resulted in the interbedding of fluvial (Selmo Formation) and shallow marine sediments (e.g. Derman 1999). Contemporanous basaltic andesite volcanism (Arger et al. 2000). the 'Yavuzeli Basalt' (Terlemez et al. 1997), also occurred, resulting in a complex succession of lava flows interbedded with the sediments (e.g. Derman 1999). K-Ar dating by Arger et al. (2000) established the age of this volcanism as c. 16 Ma, consistent with the biostratigraphic age of the associated shallow marine sediments. Before this dating study by Arger et al. (2000), this volcanism in the Pazarcik-Türkoglu area was widely thought to be Quaternary. However, the assumption of a Quaternary age was based on miscorrelation with other Quaternary volcanic fields elsewhere in SE Turkey (Terlemez et al. 1997). None the less, a Quaternary age continues to be claimed, and has been used to argue for a young (less than c. 2 Ma) age for the initiation of the East Anatolian Fault Zone. However, as dating shows that this volcanism is Miocene, such reasoning should no longer be used.

Likewise, some interpretations (e.g. those by Lyberis et al. (1992), Chorowicz et al. (1994) and Adiyaman & Chorowicz (2002)) have claimed, largely based on interpretation of satellite imagery without supporting field evidence, that the Gölbasi-Türkoglu Fault is a minor structure and the principal active deformation in its vicinity is shortening, leading to the development of the adjacent Ahir anticline and its counterparts. However, our field investigations, like those of Westaway & Arger (1996), reveal no evidence that these anticlines are active.

Their folding predates the East Anatolian Fault Zone, justifying the use by Westaway & Arger (1996) and in the present study of offset anticline axes as piercing points to quantify the young strike-slip. Our fieldwork also indicates that the interpretation by Adiyaman & Chorowicz (2002), that the Gölbasi-Türkoglu Fault is a normal fault with downthrow to the SSE, is incorrect, as no systematic component of vertical slip can be identified. In some places the topography across this fault is lower on its SSE side, but elsewhere the lower ground is on its NNW side; these local patterns can be readily explained by juxtaposition of different rock types, some more easily erodable than others, as a result of the active left-lateral slip.

The final stratigraphie unit, the 'Harabe Formation' of Terlemez et al. (1997), consists of fluvial gravel, sand and silt. This has been assigned a nominal 'Pliocene' age, as parts of it stratigraphically overlie the Yavuzeli Basalt. However, on the basis of this limited constraint, individual fluvial deposits grouped using this term could have any age from Late Miocene to Mid-Pleistocene. By analogy with the similarly designated 'Asartepe Formation' in western Turkey, it may well consist of a complex mixture of stacked fluvial units and inset deposits, forming high river terraces, such that it should arguably not be regarded as a 'Formation' in any meaningful stratigraphie sense.

Fault Zone Morphology

The Gölbasi-Türkoglu Fault can be readily identified for most of its length from the geomorphology. Its eastern end was described in detail by Westaway & Arger (1996). To the best of our knowledge, the only previous detailed investigation of the central and western Golbasi-Turkoglu Fault has been by Yalcin (1979). In the western Golbasi Basin the Golbasi-Turkoglu Fault strikes SW and runs for c. 10 km along the base of a c. 100 m high, NW-facing escarpment on its Arabian side, past Kosuklu, Catalagac and Kücükören. This escarpment exposes the uppermost Maastrichtian-Palaeocene clayey limestone of the Germav Formation; its base forms the alluvial plain in the basin interior at c. 880 m above sea level (a.s.l.), whereas its top truncates a low-relief land surface formed in the Germav Formation, which in places slightly exceeds 1000 m a.s.l.

The Kisik Drainage Catchment

About 1 km from the western end of the Gölbasi Basin, the River Aksu exits it through a gorge in this escarpment. At the entrance to this gorge (at [CB 6190 6725]) the Aksu is joined by the Kisik, one of its principal tributaries, draining an area of c. 40 km^2. From this confluence the Golbasi-Türkoglu Fault follows the Kisik upstream, passing between Sakarkaya and Soku, with a typical S60°W trend. The local relief is much more dramatic than farther NE, the escarpment on the Turkish side of the Gölbasi-Türkoglu Fault rising from c. 900 m to c. 1300 m a.s.l. as it truncates the axis of the Ahir anticline in Midyat Group limestone. West-away & Arger (1996) located this anticline axis at [CB 5950 6560] and regarded it as conjugate to the axis of the Körkün anticline on the Arabian side of the Gölbasi-Türkoglu Fault near the eastern end of the Gölbasi Basin.

The 33 km offset between these piercing points indicates the total slip on all Late Cenozoic left-lateral strike-slip faults that pass through the Gölbasi Basin, comprising the Gölbasi-Türkoglu Fault and the

NNE-SSW-striking fault zone that splays southward from it near the eastern end of this basin, which is not investigated in the present study. In contrast, the escarpment on the Arabian side of the Gölbasi-Türkoglu Fault maintains a roughly constant c. 120 m height, for instance rising from the c. 920 m river level to c. 1040 m a.s.l. where the Kisik River joins the fault line.

However, moving WSW, the exposure in the face of this escarpment passes down-section, from the Germav Formation into the underlying Besni Formation, recognizable from its distinctive coloration, and then into the Hatay Ophiolite. Tolun & Erentoz (1962) mapped the eastern end of outcrop of ophiolite on the Arabian side of the Golbasi-Türkoglu Fault in the vicinity of [PC 5640 6375], c. 1 km WSW of the point where the Kisik River reaches the fault line. East of this point, beyond the gorge flank on its Arabian side, outliers of the Besni Formation rise to c. 1100 m a.s.l.

The roughness of the local topography raises the possibility that these might overlie ophiolite, thus concealed beneath scree deposits, which may thus persist as far east as Y'. Furthermore, ophiolite is also present farther ENE in the escarpment face at least as far as [CB 5900 6550]. We return to the issue of how to determine precise piercing points in this ophiolite during discussion of the total slip on the Gölbasi-Türkoglu Fault. At the WSW end of the offset reach of the Kisik River, an offset reach of the C1narcik tributary follows the line of the Gölbasi-Türkoglu Fault for c. 2 km.

However, this is a minor drainage system compared with the Kisik trunk channel, draining only c. 3 km^2 of land area, and has evidently thus been unable to incise in pace with the regional uplift that has occurred. At its confluence with the Kisik it forms a hanging valley, where it cascades down to the level of the Kisik.

The Koca Drainage Catchment

Beyond the sector where the Cinarcik follows the line of the Gölbasi-Türkoglu Fault, this fault crosses the col, at c. 1240 m a.s.l., near Karaagac Tepe hill, entering the Koca river system. In this reach this fault typically trends S65°W, the outcrop being ophiolite on its Arabian side and partially lithified Miocene fluvial sand and gravel of the Selmo Formation on its Turkish side. Although the regional topographic gradient is southward, the more erosion-resistant ophiolite means that locally the land surface is typically higher in the immediate vicinity of the fault on its Arabian side.

The Gölbasi-Türkoglu Fault descends to the level of the Koca River and its principal tributary, the Sincer, along a gulley, passing directly south of Karaagac village and crossing the Velikler tributary, which is not significantly offset from the modern outlet gorge of the Sincer across the fault, reaching the Sincer at c. 905 m a.s.l. It then runs upstream along the Sincer, which is offset left-laterally by c. 800 m, forming a short section of linear valley between ophiolite and Selmo Formation sand and gravel.

Around the Golbasi-Türkoglu Fault leaves the Sincer River, passing for c. 1.2 km through a densely forested interfluve area, still forming the contact between ophiolite and Selmo Formation deposits before joining the Koca River at [CB 5270 6160]. Downstream of this point, the Koca flows within a gorge in the ophiolite on the Arabian side of this fault. This river is locally at c. 930 m a.s.l.; in the next c. 3 km upstream, through Kocadere, it follows the fault, rising to c. 980 m a.s.l. The land surface in the ophiolite on the Arabian side is locally at a relatively uniform altitude of c. 1070-1090m a.s.l. However, on the Turkish side, the landscape becomes progressively more subdued; at the downstream end of this reach the Selmo Formation is found at up to c. 1100 m a.s.l. whereas at its upstream end these deposits rise just a few tens of metres above river level.

Immediately WSW of the point where the Koca River joins it, at the Gölbasi-Türkoglu Fault enters a sector of very different relief, forming the contact between Midyat Group limestone on the Turkish side (Kardoga Tepe, rising to 1129 m a.s.l.), still with ophiolite on the Arabian side. The land surface in the ophiolite slopes gently westward from this point from c. 1070 m to c. 980 m a.s.l in c. 2.5 km distance, before the more deeply incised gorge system of the Gök River is reached; thus, the highest topography is now on the Turkish side of the fault.

The resulting col along the fault, at c. 1040 m a.s.l., is a distinctive landmark; dry valleys lead away from it both ENE towards the Koca and WSW towards the Gok. Westaway & Arger (1996) deduced that this localised Kardoga Tepe-Ahlankavagi Tepe outcrop of Midyat Group limestone. The c. 33 km offset between these piercing points confirms the total left-lateral slip through the Gölbasi Basin.

The Gok Drainage Catchment

The outlet of the Gök River from the Gölbasi-Türkoglu Fault, c. 1 km south of Kocalar, is a dramatic c. 120 m deep gorge, cut into ophiolite down to the c. 860 m a.s.l. river level. About 1 km farther east, truncated by the fault around, is an unnamed dry (or underfit) valley, c. 40 m deep, situated between the hills Tavsan Tepe at and Magara Tepe at, which leads into the Gök c. 1.5 km farther downstream. Two c. 1 km long gulleys join the fault line from the Midyat Group limestone of Ahlankavagi Tepe around, roughly in line with this dry valley, but with no surface drainage connection evident between the two. This Tavsan Tepe dry valley was evidently a former outlet for the drainage from the Turkish side of the fault, which presumably became abandoned as a result of the juxtaposition of the Kardoga Tepe-Ahlankavagi Tepe outcrop of Midyat Group limestone across it and the associated WSW offset of its former headwaters by slip on the Gölbasi-Türkoglu Fault.

For c. 10 km farther WSW, past Kocalar, Camhca and Sürülü, the Gölbasi-Türkoglu Fault (now oriented S57°W) is marked by the gorge of the Gök River, with ophiolite on the Arabian side and, initially, clayey limestone of the Germav Formation on the Turkish side. In this sector the Gök is joined by a succession of tributaries from the Turkish side, the Mezayok, Büyük/Sicanli, Cati, Cobanpinar, Karaagac and Alanyolu rivers. The farthest upstream of these, the Alanyolu, joins the Gök headwaters at, c. 8km WSW of the modern outlet gorge. The Germav Formation is more easily erodable than the ophiolite, so, immediately upstream of the c. 120 m deep Gök outlet gorge the higher topography is on the Arabian side of the Gölbasi-Türkoglu Fault.

However, farther WSW the land surfaces in the ophiolite and Germav Formation are both at similar levels, and the gorge depth remains a roughly uniform c. 120 m until the Karaagac confluence at, where this river is at c. 1030 m a.s.l. with the surrounding land surfaces at c. 1150m a.s.l. As in localities farther west on the Arabian side of the Gölbasi-Türkoglu Fault, the exposure in this Maastrichtian-Palaeocene sequence passes westward down-section, until ophiolitc is exposed beyond the Gök gorge to the north around.

However, the mapping by Tolun & Erentöz (1962) shows ophiolite persisting eastward within the northern flank of this gorge as far as. We postpone consideration of the potential value of these data as piercing points to constrain the total slip on the Gölbasi-Türkoglu Fault to later discussion.

The western Golbasi-Türkoglu Fault. For c. 10 km beyond the western margin of the Gök catchment, the Gölbasi-Türkoglu Fault is confined within the Hatay ophiolite, with a typical local trend of S78°W. There is little fluvial incision near the fault in this sector, and no sites where significant river offsets are clearly defined, evidently as a result of the erosion resistance of the ophiolite and the limited erosional power of these headwaters. Both the fault-line valley and the surrounding landscape reach their highest altitudes in this sector, possibly because the ophiolite is locally at its thickest.

The headwaters of the Gök drain ENE from, leaving the eastern part of an expanse of linear valley floor, Göl Alani, situated at c. 1250 m a.s.l., which forms the col between the Gök catchment and that of the Gökgecit River farther WSW. In this sector, the surrounding land surface in ophiolite reaches c. 1310 m on the Turkish side of the fault, with many hilltops at similar altitudes on the Arabian side, including Catalbükü Sirtlan, Gökgecitbasi Tepe, and an unnamed c. 1300 m summit south of Göl Alam at [CB 3845 5355].

The western end of Göl Alam is drained WSW from by the Gökgecit River, which follows the fault for c. 4 km to Kartal, leaving the fault line on its Arabian side at [CB 3415 5320]. Kartal village is located on the sloping surface of an alluvial fan on the Turkish side of the fault, which reaches as low as c. 1050 m a.s.l. in the vicinity of the Gökgecit outlet gorge, where this river has incised to c. 100 m lower. Just west of Kartal, at [CB 3405 5320], the Gölbasi-Türkoglu Fault crosses a low col in the alluvial fan surface, which forms the drainage divide between the Gökgecit and Cigli rivers. From this point, the Cigli River follows the linear valley along the fault to Türkoglu.

Around [CB 3115 5165], c. 1.5 km ENE of Cigli (or Ayanusagi) village, the fault ceases to be confined within ophiolite on its Arabian side. For c. 2 km from this point it passes through an area mapped by Terlemez el al. (1997) as an outcrop of the 'Harabe Formation', but which seems to be more appropriately described as another large alluvial fan, shed from the ophiolite to the north, through which protrude inliers of ophiolite and Besni Formation sediment. This locality, where ophiolite ceases to be juxtaposed on the Arabian side of the fault, was used by Westaway & Arger (1996) as a piercing point to infer 16 km of total slip, relative to the western end of the ophiolite outcrop north of the fault at Türkoglu. However, as with other possible piercing points related to margins of ophiolite outcrops, the precise location of this one is difficult to establish: it could arguably be anywhere from [CB 3115 5165] (noted above) to [CB 2950 5100], representing a projection onto the fault of the westernmost outcrop of ophiolite, in the vicinity of Cigli village. Beyond Cigli, the Gölbasi-Türkoglu Fault can be traced near the foot of the ophiolite escarpment on its Turkish side, against the Quaternary alluvial plain of the Cigli River. At [CB 2680 5080], c. 3 km WSW of Cigli, where transected by the road from Kahraman Maras to Gaziantep, this plain is c. 600 m a.s.l. and the adjacent ophiolite rises to c. 700 m a.s.l. Moving WSW, the surface to the plain descends gradually to c. 470 m a.s.l. at [CB 1450 4250] near Türkoglu, where the Aksu River crosses the fault from south to north. In contrast, the ophiolite rises to 903 m a.s.l. at the summit of Koroglu Tepe [CB 1990 4975] before gradually descending to beneath the Quaternary alluvial plain around Turkoglu.

The Gölbasi-Turkoglu Fault can be clearly traced near the base of this escarpment past Tevekkelli, Kocalar and Öksüzlü, with a typical trend of S62°W, a clear fault scarp being evident for c. 2 km around Tevekkelli between [CB 2515 4950] and [CB 2360 4870]. Pervasive cover by Pleistocene slope deposits conceals the ophiolite west of c. [CB 1770 4550], near Turkoglu. However, directly downstream of the line of this fault the Aksu at c. 470 m a.s.l. is flanked by bluffs rising to c. 540-550 m a.s.l., around [CB 1550 4450] in its right bank and as far west as [CB 1300 4300] in its left bank, which may represent the westernmost part of this ophiolite unit, obscured beneath Pleistocene deposits. As with other sites already discussed, use of this margin of the ophiolite as a piercing point for estimating the total slip on the Golbasi-Turkoglu Fault is thus somewhat problematical.

Late Holocene Landscape Evolution

Archaeological excavations at the coast of A'asu, in Tutuila Island of American Samoa, exposed a depositional sequence spanning the past circa 700 years. With the period represented, sedimentation rates exceeded 10.15 cm per century in the valley floor and 16.34 cm per century along the valley margin. The occupational history may correlate with changes in climate, sea level, and coastal geomorphology. Although the evidence accords with the expected responses to the Little Climatic Optimum (circa 1050 to 690 B.P.) and Little Ice Age (circa 575 to 150 B.P.), the most plausible explanation for the A'asu case is that environmental change accompanied expansion of upland land use. Based on evidence here and elsewhere in Tutuila, it is proposed that the establishment of fortifications, monuments and permanent settlements in the uplands was part of a broader pattern of land-use expansion beginning in the fourteenth century A.D. Thirty-five years ago, a study of oxygen-isotope ratios from a speleothem (cave formation) in New Zealand demonstrated that a rapid temperature drop occurred in southern Polynesia in the fourteenth century.

Wilson et al. (1979) proposed that the violent climatic conditions accompanying this temperature drop would not only have resulted in increased latitudinal temperature variants but would also have had catastrophic effects on agriculture. At the same time, the more violent climate would have made long-distance voyages inherently more dangerous. More recently, Bridgeman (1983), based on an extensive review of previously published climatological data, proposed that climatic change may have contributed to a general collapse in Polynesian migrations after A.D. 1350. These studies have received renewed discussion since 1995 when Nunn (1995, 1998, 1999, 2000) published sea level data synchronised to climate change, proposing his own theory that changing climatic conditions had a profound regional effect on Polynesian culture.

A better understanding of climate-induced landscape change is necessary in order to better model human response to dynamic ecosystems. Climate change may result in systematic environmental adjustments in an alluvial catchment. Because archaeological sites are found within Holocene-age sediments, it is necessary to understand the late Quaternary history of the region in order to understand prehistoric settlement patterns and the completeness of the archaeological record.

In order to determine the extent to which changes in the late Holocene geological record correlate with climate and ecological changes over the same period, geoarchaeological explorations were undertaken in A'asu on the remote northern shores of Tutuila Island, American Samoa. As the geochronology became known, it was apparent that the sedimentary record had been highly irregular in the late Holocene, particularly after A.D. 1300. Explaining the causes of that dynamism became the focus of research over the next several years. In this chapter, geoarchaeological data from A'asu are compared with similar data from coastal settings and combined with recent research on the establishment of mountain settlements Pearl 2004) to show that changes in climate, settlement patterns, and landscape evolution converged, beginning in the fourteenth century A.D., a time when social complexity was on the rise in Samoa. The lush valley of A'asu is situated on the north coast of western Tutuila, with numerous inhabitable valleys undisturbed by roads and contemporary settlements, standing in stark contrast to other more populated and developed areas of Tutuila. The valley floor extends about 0.5 km inland and comprises about 0.1 [km^2] of mostly gently sloping terrain. A considerable freshwater stream, derived from runo off and several springs high in the mountains, dissects the valley floor. The springs provide a perennial source of fresh water and likely made this valley particularly attractive to prehistoric inhabitants.

A'asu provided a fortuitous starting point for geoarchaeological study because its constricted valley opening yet fairly large catchment cause it to act like a "graduated cylinder," accentuating periods of erosion and stability in the stratigraphic record. The stratigraphy of A'asu Valley is seen as a component piece in the larger picture of late Holocene landscape evolution in eastern Samoa.

The valley is formed of alluvial sediments transported along A'asu Stream and its tributaries and colluvium eroded from the steep-sided valley walls. Slope-eroded colluvium has created thick deposits along the valley floors and margins that interfinger with alluvium deposited by the stream. Along the coastal plain, sediments have been reworked by wave action, storms, and fluctuations of sea level.

A village at the mouth of the valley was encountered in 1787 when the French explorer La Perouse made landfall on Tutuila. At their meeting in 1787, Samoans and French explorers clashed at A'asu, resulting in the deaths of an undetermined number of Samoans, the French expedition's second in command (de Langle), and 11 other members of the landing party. Today there are no permanent inhabitants of the village, but it is a popular site for seasonal fishing activities. Taro and other cultigens are farmed nearby.

Emerged reef fragments visible from A'asu and at more than 20 other locations around the island indicate that relative sea levels are lower at present than earlier in the Holocene. Stearns (1944) was the first to consider that these were evidence for a higher sea stand, but their late Holocene dates were not known until recently. Nunn (1998) dated six reef fragments ranging from 0.75 to 2.11 m above mean sea level, each of which dated to the last 1000 years.

Similar data have been obtained in Western Samoa, where raised beach rock indicates a mid-Holocene highstand of 0.8-2.3 m at 1200-1850 [+ or -] 70 years B.P. Independently, Dickinson (2001) reports the differential elevation between modern and emergent wavecut coastal benches offshore of Tutuila (on the islet of Aunu'u) at 1.8 [+

or -] .01 m, a figure very close to the calculated theoretical value of 1.9 m. Furthermore, coastal progradation on 'Upolu since 700-1000 years B.P. has been attributed to systemic adjustments caused by lowering sea levels. This interpretation underscores the need for additional studies on the impact of climate and sea level change on landscape evolution in Samoa. Though some islands in the Samoan chain are experiencing subsidence or upflexure due to volcanic activity, the islands of Tutuila and Aunu'u are not thought to have experienced any significant vertical change.

Eustatic shoreline movements on Tutuila are key to interpreting the geomorphic changes at A'asu. Comparing an elevated beach rock geochronology with data from 18 other tectonically stable sites in the Pacific, Nunn (1998) summarised eustatic shoreline movements on Tutuila during the last millennium as follows: (a) a sea level rise, coincident with a period of warming known as the Little Climatic Optimum, between 1050 and 690 B.P.; (b) a sea level drop during the Little Ice Age, between 575 and 150 B.P.; and (c) a period of recent warming during the last 150 years. The transition between Little Climatic Optimum and Little Ice Age, which peaked between 690 and 575 B.P., was marked by rapid cooling and possibly a concurrent increase in rainfall.

A study of the geochronology of A'asu was conducted simultaneously with archaeological investigations; consequently, geological profiles discussed herein are those of the archaeological excavation units (sondages). The work was completed in two eight-week field seasons in the months of May and June of 2001 and 2002. The research was initiated by the author as the first step in a multiyear research plan to better understand the late precontact archaeology of Samoa. The fieldwork was conducted by the author, several graduate and undergraduate students, and professional excavators. Because the final report of the archaeological investigations is still in preparation, the general methods and procedures are summarised below.

The archaeological plan called for multiple 1 x 1 m sondages, sometimes contiguously grouped into blocks, to be located across the landscape. Fifteen such sondages were tested in this manner and labelled blocks A through E. Of those blocks, this study will focus only on blocks A and D, which provided deep exposures of alluvium exhibiting stratification and abundant charcoal for dating (the remaining blocks focused principally on surface deposits). Block A was situated inland near the valley margin in a setting that was expected to have abundant colluvium, while block D was situated on the modern floodplain of A'asu Stream. Both of the blocks were located in depositional settings, but it was anticipated that block A would reveal higher sedimentation rates due to its location at the toe of a steep valley wall. Ultimately, block A was expanded to include two adjacent sondages, reaching a maximum depth of 2.0 m. Block D was expanded to include five contiguous sondages, reaching a maximum depth of 1.3 m. Excavations proceeded in arbitrary 10-cm levels or by natural stratigraphy when change occurred within these. All sediments from block A (excavated in year 1) were screened through 1/8" mesh, while 1/4" mesh was used in block D (excavated in year 2). The change was made between seasons after careful consideration of the materials being recovered and time available to us in the second season. Water screening was employed at block A when the clay fraction exceeded about 30 percent. Artifacts were point-plotted before removal when possible. A stratigraphic cross-section was made for each completed excavation block, which was then photographed and backfilled.

Sediment samples were routinely taken at least once every 10 cm. Charcoal was so ubiquitous that no fewer than four samples were taken every 10 cm. The process of washing, sorting, cataloging, and analysing artifacts began in the field and was completed in the archaeology laboratory at Texas A&M-Galveston. Soils and sediments exposed in sections were described using standard procedures and terminology outlined by Soil Survey Staff (2003). Major lithostratigraphic distinctions are made by grouping substrata into major strata, numbered consecutively I, II, and III. Subtle changes within these strata are given an alphanumeric designation.

For interpretive purposes, the conventional radiocarbon age was calibrated with the OxCal 3.10 radiocarbon calibration software using the INTCAL04 atmospheric carbon curve for calibration.

Stratigraphic Sequence

Block A

The two sondages comprising block A were situated directly on a stone-lined house platform. Such platforms were the predominant foundation type in ethnohistoric accounts of Samoa, providing foundations for dwellings, ceremonial guest houses, cooking houses, and even churches and other buildings. This one, like most others, is covered in a pavement of gravels ('ili'ili) and bordered by one course of basalt stones.

Most of the sediments encountered were terrigenous. However, some sands, corals, and shells were intermixed with the alluvium and colluvium, especially in the uppermost strata. Soil horizons (pedostratigraphy) and lithological discontinuities (lithostratigraphy) were both noted. Based on their lithology and depositional features, ten divisions were recognised within the three major strata.

Strata Ia-Ie—Strata Ia through Ie re.ect the construction sequence of the house platform, with Ie consisting of loose coral and basalt cobbles representing the .rst stage of construction. The contact between strata Ie and IIa beneath is abrupt and irregular, indicating that some erosion or intentional clearing may have preceded the deposition of Ie.

A conventional radiocarbon determination of 100 [+ or -] 1.4 pMC (A-12406) was obtained on charcoal from the bottom of stratum Ie. Taken at face value, 100 percent modern carbon would indicate a date of about A.D. 1950. However, the lower limit (2-sigma) calculates to the interval A.D. 1660-present. The upper limit falls in the mid-1950s if all the carbon was fixed in a single year. Better precision is not possible in this interval of time because of high atmospheric carbon variations. The corrected age for this sample is equivalent to 0 [+ or -] 110 B.P. Most likely the structure dates to A.D. 1800-1930, although an earlier date cannot be ruled out.

Strata IIa-IIb—Stratum II consists predominately of clay loams, but a significant fraction of pebbles, gravels, and shells make the group quite heterogeneous. Stratum IIa is made of very dark grayish-brown sandy clay loam, with as much as 25 percent of the matrix pebble and larger-sized clasts. Most pebbles are subrounded, indicating not only their colluvial origin but that they have rolled some distance prior to deposition. There are abundant marine shells throughout these sediments, many exhibiting burning, and high densities of other cultural artifacts, though their numbers decrease with depth. Abundant charcoal and bone were also recovered at the contact with the overlying stratum. Stratum IIb transitions smoothly over a 10-cm boundary. This is a pedogenic distinction; that is, the boundary was formed through natural soil-forming processes. It is identical in colour and texture to IIa and is distinguished principally by soil structure and the rapid reduction in the number of marine shells, though they still constitute a small fraction of the horizon.

A radiocarbon age of 355 [+ or -] 35 B.P. (AA-51255) indicates that deposition of stratum II was under way in the fifteenth century. It is separated from stratum III by an abrupt irregular boundary, an indication that erosion occurred between the two. Strata IIIa-IIIb—Stratum III consists of fairly homogenous sandy clays, with clay content slightly increasing with depth. The principle differentiation between sub-horizons a and b is that the latter is slightly darker (dark brown vs. brown) and has higher clay content (sandy clay vs. sandy clay loam). The changes in colour and texture are related and occur gradually with depth. Stratum III is generally more fine grained than the overlying strata, suggesting, in the absence of any evidence for pedogenically altered clays, that water transport played a greater role in depositing the sediments. However, occasional angular rocks and a large boulder encountered at the base of block A reveal that much of the deposition is still colluvial (from the slopes) rather than fluvial (from the flooding of A'asu Stream). Stratum IIIb is the deepest and oldest encountered, and there is abundant charcoal dispersed in this layer.

Four radiocarbon ages were determined on charcoal from stratum III. Two samples were taken from the burned hearth mentioned above. A third and fourth sample were taken from dispersed charcoal beneath the hearth area elsewhere in the layer. The samples were separated by a distance of no more than 20 cm. The two samples from the hearth returned calibrated radiocarbon ages of 625 [+ or -] 35 B.P. (AA-51257) and 630 [+ or -] 40 B.P. (BETA-171844). The samples beneath the hearth returned calibrated radiocarbon ages of 635 [+ or -] 35 B.P. (AA-51256) and 710 [+ or -] 40 B.P. (BETA-171845).

Block D

The five sondages comprising block D were situated on the alluvial plain. At first glance it appears to be an area devoid of architectural features, and initially the first sondage was placed there to establish the underlying stratigraphy of the alluvial plain. However, archaeological discoveries necessitated expanding the block to a full 5 [m^2]. Cultural materials were encountered in each stratum, including an adult human burial in IIIa. Based on their lithology and depositional features, seven divisions were recognised within the three major strata. Like block A, most of the sediments encountered were terrigenous. However, some marine sands, corals, and shells were intermixed with the alluvium and colluvium, especially in the uppermost strata, such as Ia below.

Stratum Ia-Ic—A 5-cm veneer of sand coats the surface. This sand, comprising stratum Ia, likely resulted from hurricanes that blasted this coastline in the 1990s. Former residents report that the village was leveled at that time and significant volumes of seawater flooded the low-lying parts of the village.

Stratum Ia abruptly overlies stratum Ib, a very dark grayish-brown sandy loam. Pebbles and gravels make up to 20 percent of the matrix, which also contains small amounts of fire-cracked rock and charcoal.

Stratum Ic represents a single anthropogenic horizon. Its coral cobbles and matrix of gravelly, sandy clay loam was initially encountered in the floor of some sondages. As the excavation area was laterally expanded, it became clear that this clast-supported matrix was bounded on the south by a line of basalt boulders. These were 20-30 cm across and are consistent with those seen bordering domestic areas in ethnohistoric contexts. The floor was punctuated by a cobble-lined pit feature extending into the underlying strata.

Strata IIa-IIb - Directly beneath the contact of stratum I, a soil-enriched, very dark grayish-brown sandy to gravelly clay loam (up to 20 percent pebbles and gravels) was encountered (stratum IIa). Many marine shells were also encountered, as well as some bone, charcoal, and other cultural material. This stratum, distinctly darker than the overlying layer, could be seen clearly following the outline of pit features in adjacent sondages. Charcoal from the base of IIa provided a radiocarbon age of 320G35 B.P.

Stratum IIb transitions smoothly over a 10-cm boundary. It is distinguished from IIIa by a slightly lighter value (dark grayish-brown), as well as by decreasing gravels, shells, and cultural finds. It terminates abruptly and lacks any soil development. Charcoal from the base of this stratum provided a radiocarbon age of 340G50 B.P.

Strata IIIa-IIIb—Stratum III consists of homogenous sandy clays, with clay content slightly increasing with depth. It is lithologically distinctive from stratum II; the abrupt planar boundary between them represents an erosional disconformity. The principle differentiation between IIIa and IIIb is that the former is strongly discolored by abundant charcoal. The entire horizon can be characterised as an ashy mass of poorly sorted, decomposing charcoal fragments in sandy clay, with some lithics and faunal remains. It would not be appropriate to refer to the layer as a lens because its lateral extent remains unknown. Further, there was no evidence of any depression or depositional cavity that might have bounded the layer. The lower boundary was clear but not abrupt, with a transition to IIIb visible over about 5 cm. Charcoal content in the underlying stratum was dramatically lower.

Stratum III is generally more fine grained than the overlying strata, suggesting that fluvial transport played a greater role in depositing the sediments. Stratum IIIb is the deepest and oldest of the sedimentary units. It contained abundant cultural material as well, including an adult burial, a fishing weight (not associated with the burial), and other cultural materials. A single radiocarbon age was determined on charcoal from stratum III. A large sample of concentrated charcoal from IIIa returned an AMS age of 650G50 B.P.

Interpreted Geochronology of A'ASU

Correlation of the stratigraphy between blocks A and D is based on the lithology and soil morphology, and major strata given the same designation (i.e., I, II, or III) are interpreted as being part of the same depositional layer. Atmospheric radiocarbon fluctuations between A.D. 1325 and 1375 make calibration of dates in the fourteenth century problematic. Conventional radiocarbon ages between about 500 and 700 B.P. have a bimodal probability distribution when calibrated. Furthermore, many of the dates that I cite were calibrated using different calibration curves than are now available. Consequently, for comparative purposes I will emphasise the ages in radiocarbon years (B.P.), which are consistently reported in the papers cited and are independent of calibration curves.

These dates essentially cluster into three groups. A chi-square of the five dates from stratum III reveals that they are statistically the same at the 95 percent confidence interval (648 [+ or -] 17; df = 4; T = 3.2; [chi square] (.05) = 9.5). Similarly, a chi-square of the three dates from stratum II reveals that they are statistically identical at the 95 percent confidence interval (341 [+ or -] 24; df = 2; T = 0.4; [chi square] (.05) =4 6.0). This gives a high degree of confidence that the gap between stratum II and III is real and suggests that the vacuity represents about 300 years.

The oldest radiocarbon date from A'asu is 710 [+ or -] 40 years B.P., obtained from charcoal in stratum IIIb in block A. A small number of lithics, including an adze fragment, were recovered beneath this charcoal sample. Utilisation of A'asu Valley therefore began prior to 710G40 B.P. During the Little Climatic Optimum (~1050 B.P. to ~690 B.P.), the predominance of well-sorted, fine-grained alluvium in the valley suggests that fluvial processes predominated. The dispersed charcoal in the deepest sondages probably indicates that swidden agriculture played an important role in subsistence patterns at that time, though .re from natural causes cannot be ruled out. It should be noted, however, that while human use of fire for agricultural purposes in moist tropical environments is very common, large natural fires are very uncommon.

Around 650 [+ or -] 50 B.P., the valley floor was blanketed by charcoal, at least locally, after which the surface was truncated by erosion. The precise timing of this erosional event is unknown because the sediments related to it have been flushed out of the valley. The direct source of the charcoal is also undetermined, but erosion across the valley floor may be due to decreased vegetation cover. A date on the overlying stratum of 355 [+ or -] 55 B.P. means that the lacuna represented by the erosion includes the transitional period between the Little Climatic Optimum and the Little Ice Age of the Late Holocene (~690 to ~575 B.P.) and might be directly correlated to it.

Deposition of the stratum II was under way no later than 355 [+ or -] 35 B.P. On the valley margin, rapid colluvial deposition predominated, with minimal alluvium, while deposition on the valley floor was a mixture of colluvium and alluvium. The increased sediment yield was coincident with the Little Ice Age, when both temperature and sea level were lower than present. A mean rate of sedimentary deposition was calculated by dividing minimum vertical accretion (depth between the bottom of stratum I and the deepest radiocarbon determination in each block) by mean accumulation time (as determined by the radiocarbon intercept).

Due to the contribution from the steep valley walls, rates of sedimentary deposition along the valley margins were 62 percent higher than on the valley floor. Along the valley margins, sediments accumulated at a mean rate of 16.34 cm per century (116 cm over 710 years). Of course, because of the lacuna left by erosion, the actual rate of deposition was higher at times. Along the floor of the valley, sediments accumulated at a mean rate of 10.15 cm per century (66 cm over 650 years). While perhaps over-generalised, these calculations indicate that deposition has been significant over the last 700 years.

Throughout the Holocene, sea level constrained the movement of sediment towards the sea. During the Little Climatic Optimum, the alluvial plain aggraded. Then, as the climate transitioned to the Little

Ice Age, A'asu Stream cut into its alluvial plain and its gradient increased as it adjusted to a lower sea level. Sea level rise over the past 150 years or so has begun to stabilise the alluvial plain. In Tutuila and Aunu'u, the drop in sea level after the mid-Holocene high-stand exposed more landmass and created conditions for prograding shorelines.

The aceramic site of A'asu has a rich historic and prehistoric archaeological record. So that readers may better understand the archaeological context, a brief summary of the archaeology is provided here. The modern surface is a palimpsest of recent and ethnohistoric discard. Secondary maintenance has removed most artifacts from major walkways, but discarded materials are found around many of the house foundations and at the periphery of the village. Discards range from wire, glass, and metal "trash" to the occasional polished stone flake or adze fragment. Most of the existing house foundations are cement slabs of recent origin, though there are several traditional stone fale foundations as well.

The archaeology of the uppermost excavation layers (stratum I) is closely related to recent village activities. Modern and historic glass, metals, and plastics were recovered, including such diagnostic objects as nails, buttons, and religious devotional medals. These artifacts are a clear indication that the site was well connected with long-distance trade networks in historic times. Many of the artifacts are marine related-boat-building nails (copper with roves), shell-motif beads, and shell buttons-attesting to the strong connection that the people of A'asu had with the sea.

The archaeology of stratum II was increasingly prehistoric in character, consisting principally of lithics and faunal remains that declined with depth, an indication of the poor preservation conditions at the site. The lithic artifactual remains, however, showed a bias towards non-adze stone tools. A number of worked flakes, especially unifacial scrapers, were encountered, as well as a small number of adzes. Use-wear indicates that the adzes were usable finished products, but they were not highly polished like many other Samoan adzes. Stratum II contained a limited number of historic artifacts, including a kaolin clay pipe fragment of a style that was common at the end of the eighteenth century.

The exclusively prehistoric archaeology of stratum III was very similar to the archaeology of stratum II. Indeed, as of this publication I am not prepared to say that there was a significant difference in terms of lithic technology, and even fewer faunal remains were present. An oval stone “weight” was recovered that might have been a fishing or net weight. A burial was also recorded in block D, stratum IIIb. No cultural materials were associated with the burial, which was lying on its right side—facing north, towards the sea—in a flexed position.

Cultural Stratigraphy and Site Formation Processes

Three archaeological components are present at Masterov Kliuch. Component I is an early Upper Palaeolithic occupation occurring from 90-100 cm below the surface within geologic unit 2. Component II also appears to be an early Upper Palaeolithic occupation; it occurs within unit 4 at a depth of 30-60 cm below the surface. Component III is a Bronze Age occupation ranging in depth from 0-20 cm below the surface within units 5 and 6. Given the complex geologic context of the Masterov Kliuch site, an important part of our research has been to establish the integrity of the site’s Palaeolithic components, especially in terms of natural site deformation processes related to colluviation and cryoturbation.

Three indicators of site integrity—vertical distribution of artifacts, horizontal distribution of artifacts, and presence of conjoined artifacts—were studied in order to ascertain the degree of disturbance by colluvial processes. For components I and III, vertical distribution of artifacts is relatively tight, with component I occurring within a 10-cm-thick band and component III occurring within a 12-cm-thick band. A similar pattern can be seen in the horizontal distribution of artifacts from components I and III, with artifacts being situated in identifiable clusters across the excavation. Further, 13 artifacts from component I were conjoined; the average horizontal distance between these conjoined artifacts is 20.25 cm, and the average vertical distance is only 1 cm. The tight vertical and horizontal distributions of artifacts, as well as the close horizontal and vertical proximity of conjoined pieces, suggest that the artifacts of component I lie in a primary context.

Component II artifacts, however, have a much greater vertical distribution than those in components I and III (component II has an average thickness of about 20 cm), and the horizontal distribution of artifacts appears more scattered than in components I and III. Further, no artifacts from component II could be conjoined. These data suggest that component II is redeposited. Frost-heaving (the movement of artifacts due to repeated freezing and thawing of sediment) is also a factor affecting northern archaeological sites. To evaluate the degree to which freeze-thaw processes impacted the cultural components at Masterov Kliuch, we measured trend and plunge of all large-sized artifacts encountered in situ in the Palaeolithic components, using a Brunton pocket transit. Twenty-two such artifacts were analysed in this way—16 for component I and 5 for component II.

Plunge measurements show that roughly half of the artifacts lie within 45 [degrees] of horizontal; that is, they lie more flat than upright. The other half lie more vertically upright, with plunge measures of between 45 [degrees] and 90 [degrees]. Of these, only three artifacts have plunge measurements of 90 [degrees]. Once reaching 90 [degrees] plunge, artifacts tend to move upward through the profile. Thus, although frozen ground processes appear to have reoriented some artifacts, there is little indication that they have displaced them vertically. Trend measurements, further, show no obvious pattern in the direction that the artifacts plunge, and few actually are trending along the slope of the site (about 100 [degrees] east of north), suggesting that artifacts of component I have been reoriented by minimal frost-action, but probably not slumping or slopewash.

The stratigraphic and provenience information from Masterov Kliuch show that while both Palaeolithic components lie in colluvial deposits, only component II is redeposited. Component I appears to lie in its primary place of deposition. Although frost-heaving has affected the orientations of the artifacts from components I and II, this process does not appear to have affected the locations of these artifacts.

Radiocarbon Chronology and Age of Cultural Components

Samples of bone (n = 3), tooth enamel (n = 1), charcoal (n = 1), and soil organics (n = 3) from the geological units and cultural components at Masterov Kliuch were dated through accelerator radiocarbon (AMS^{14}C) procedures. Charcoal was not well preserved in the site, occurring only in the uppermost unit in association with archaeological component III. For this reason, we concentrated on the dating of bone and other materials. When appropriately purified using XAD-2 resin, bone with significant amounts of intact collagen (typically greater than 5 percent of original amount of protein) can provide accurate age estimates (Taylor 1997). Pretreatment and AMS ^{14}C analysis of all samples was conducted at the NSF-Arizona AMS Facility, following standard methods described by Long et al. (1989) and Jull et al. (1983) for the AMS ^{14}C dating of bone and charcoal, respectively. All dates are reported as uncalibrated.

Five radiocarbon ages were obtained from stratigraphic unit 2. Three samples of bone, one sample of tooth enamel, and one sample of organic matter were dated. The most reliable ages were derived on the three bone samples. Two of the bones were collected from archaeological component I during our excavations in 1996. These XAD-purified samples of bone (AA-23640 and AA-23641) had relatively high amounts of original protein (11.3 and 14.8 percent, respectively) and yielded ages of 32,510 [+ or -] 1440 and 29,860 [+ or -] 1000 B.P., which overlap at two-sigma. The other XAD-purified bone sample (AA-8888), which yielded an age of 24,360 [+ or -] 270 B.P., was collected from Meshscherin's 1991 excavation (Goebel 1993). This sample was collected from stratigraphic unit 2, but above archaeological component I and may be from a later brief occupation. These dates clearly indicate a pre-Sartan (pre-late glacial) age for unit 2 and that archaeological component I dates to roughly 30,000 B.P.

A pre-Sartan age for unit 2 is also supported by the frost cracks that extend from the top of this unit as well as the absence of frost cracking in overlying units. These frost cracks probably developed during the Sartan glacial period, as at other Upper Palaeolithic sites in the Baikal region (Bazarov et al. 1982; Tseitlin 1979). Thus, unit 2 and its associated archaeological component must pre-date the Sartan glacial period based on geologic evidence. This geologic scenario is supported by a radiocarbon age on organic-rich sands found within a channel overlying the frost-cracked surface of unit 2 in a test pit 50 m to the northwest of the main 1996 excavation area. Organic matter from this sand yielded an age of 18,850 [+ or -] 135 B.P. (AA-23647). This organic material appears to have been derived from the erosion of soils that had developed on the slopes above the site.

Two aberrant ages were obtained from stratigraphic unit 2. One small fragment of tooth enamel from archaeological component I yielded an age of 19,415 [+ or -] 260 B.P. (AA-23642). Also, a small fragment of what was thought to be charcoal was collected near a frost crack in the sidewall exposure of Meshcherin's excavation. This sample yielded an age of 18,335 [+ or -] 320 B.P. (AA-23643). These ages are at odds with the older bone-derived ages from component I and the geologic evidence. There are several reasons why these younger ages are disregarded. The age of 19,415 [+ or -] 260 B.P. (AA-23642) was derived on the inorganic apatite fraction of the tooth. Apatite is notorious for yielding inconsistent results because a number of mechanisms can significantly alter carbon-isotope values in the apatite structure.

The date of 18,335 [+ or -] 320 B.P. (AA-23643) turned out not to be derived from charcoal, but instead from an aggregate of organic matter. We believe this aggregate most likely represents the post-depositional movement of an organic particle into unit 2. Since this sample was collected only 2 cm from a visible crack, it may have been translocated into unit 2 from higher in the profile. As mentioned above, in some places on the site organic-rich sands dating to 18,850 B.P. are found overlying unit 2. Both the date on organic matter from unit 2 and the date from the overlying organic-rich sand (in the nearby test pit) are statistically indistinguishable at one-sigma. It seems likely that a sample of this organic-rich sand was translocated downward through the profile via a frost crack into the underlying unit 2. Thus, this age is considered invalid.

In an attempt to date geological unit 1, at the base of the profile, a bulk sample of soil organics from geologic unit 1 (taken from about 150 cm below surface) was AMS [sup.14]C dated and yielded an age of 7630 [+ or -] 65 B.P. (AA-23646). This date is clearly too young based on the overlying dates from unit 2 and can be disregarded. Stratigraphic units 3, 4, and 5, and archaeological component II are undated. Based on the organic-rich sand age and artifacts from component II, these appear to date to the late Upper Pleistocene, perhaps 18,000-10,000 B.P.

A sample of charcoal from a small hearth feature in component III near the top of the stratigraphic profile yielded an AMS [sup.14]C age of 2895 [+ or -] 45 B.P. (AA-23648), providing support for the presumed late Holocene age of this cultural component. Given these AMS [sup.14]C determinations, as well as the above review of site stratigraphy and site formation processes, we can make the following conclusions about the age of the Masterov Kliuch sediments and cultural components. Unit 1 was deposited sometime prior to 30,000 B.P. The frost cracks and small ice-wedge pseudomorphs that originate along the upper contact of unit 1 perhaps formed during the Konoshchel'e cold snap, dated elsewhere to 33,000-31,000 B.P., or during some earlier stade of the early or middle Pleniglacial. Unit 2 and component I are AMS ^{14}C dated to about 30,000 B.P., the beginning of the Lipovsko-Novoselovo interstade (independently dated to 30,000-22,000 B.P.). This is further supported by the extensive network of frost cracks and ice-wedge pseudomorphs that originate from the upper contact of unit 2; these probably formed during the height of the Sartan stade (22,000-17,000 B.P.). Unit 3 and component II have not been AMS ^{14}C dated, but, given their stratigraphic position above features relatively assigned to the last glacial maximum, as well as the platy structure of the sediment, must have been deposited (from upslope) sometime during the late glacial (17,000-10,000 B.P.). Units 4, 5, and 6 likely formed during the Holocene (10,000 B.P. to the present). Component III, found within units 5 and 6, dates to about 3000 B.P. and thus can be assigned to the Transbaikal Bronze Age.

Archaeological Assemblages and Features

The two Masterov Kliuch Palaeolithic components are described in detail below. Because it was not a focus of our study, the Bronze Age component is only briefly presented. For definitions of terms used to describe cores, tools, and other lithic artifacts, readers are referred to Andrefsky (1998) and Goebel (1993).

Component I

Cultural component I consists of a relatively dense band of lithic artifacts, with two distinct concentrations occurring in the 6-[m.sup.2] excavation, including a small cluster of flaking debris in the northwestern corner of the excavation (square 26T), and a larger cluster of retouched artifacts, cores, and flaking debris in the eastern half of the excavation (squares 25P, 26P, 25R, 26R). Within the latter cluster, two lithic technological activities are evident: (1) primary reduction activities represented by a concentration of 6 cores and about 60 cortical flakes, and (2) secondary reduction activities and tool use represented by a concentration of 18 retouched artifacts and nearly 30 retouch chips.

The few fragments of bone that were encountered during excavation of component I came from the eastern half of the excavation, in association with the concentrations of tools and retouch chips. The component I assemblage consists of 367 pieces, including 360 lithic artifacts and 7 small bone fragments. Among lithic raw materials, dark gray cryptocrystalline silicate (ccs) dominates, making up 73 percent of the assemblage. Other materials include dark red ccs (11 percent) and speckled gray ccs (10 percent), while translucent tan/gray ccs (1 percent), brown ccs (1 percent), tan ccs (1 percent), and green ccs (1 percent) occur in low frequencies. There are also four splintered stones of clear quartz and two of clear quartzite that may be manuports. All of the ccs materials are available locally in alluvium of Gyrshelun Creek and the Khilok River. The debitage assemblage

(338 pieces) includes 7 cores, 66 cortical flakes, 179 flakes, 30 blades and blade fragments, 28 retouch chips, and 28 splinters. Cortical flakes (making up 18 percent of the lithic assemblage) include 37 primary flakes, 24 secondary flakes, and 5 fragments. These are typically made on dark gray and speckled gray ccs. Cortical flakes occur on every type of raw material present in the assemblage, further supporting the notion that raw materials were obtained locally. Further, the relative frequencies of raw material types are virtually the same for cortical flakes and noncortical flakes, indicating that unworked cobbles were carried to the site for reduction.

Core preparation and flake removal techniques were relatively expedient. The seven cores are informally prepared and include two monofrontal unidirectional flake cores made on cobbles of speckled gray ccs and dark gray ccs, a bifrontal bidirectional flake core made on a dark gray ccs cobble, a small end core (blades were struck from the end of the core rather than the face) made on a dark red ccs flake, a bipolar core made on a dark gray ccs flake, and two possible core tablets (platform rejuvenation spans) on dark gray and speckled gray ccs. Platform surfaces were simply prepared, with 84 percent of all cores and their removals having smooth platforms, and 11 percent having cortical platforms. Trimming and grinding of platform edges is evident on 55 percent of debitage pieces.

Blades and blade fragments make up 9 percent of the debitage assemblage. No large blade cores, however, were encountered during our excavations in 1996, but earlier excavations by Meshcherin in 1996 did yield one obvious blade core from component I (Goebel 1993). This is a unidirectional flat-faced blade core on dark gray ccs. Among the 21 tools, 11 are made on flakes or cortical flakes, indicating expediency in the production and selection of tool blanks. Nine tools are made on blades, and one, a chopper, on a cobble.

The presence of retouch chips in the debitage assemblage indicates that some secondary reduction activities also occurred at the Masterov Kliuch site. However, as with core preparation and blank manufacture, tool resharpening appears to have been expedient. Retouch invasiveness is minimal, with 14 of 22 tool edges having flake scars that extend less than 3 mm from the tool margin. Only two artifacts have retouch scars that are greater than 10 mm; these include a cobble chopper and denticulate. The 22 tools in the assemblage include 8 retouched blades and blade fragments, 5 retouched flakes, 2 knives, 2 denticulates, 1 of each of the following: graver, notch, cobble chopper, possible burin spall, and combination tool. Six of eight retouched blades are bilaterally retouched. Both knives are cortically backed, one on dark gray ccs and the other on dark red ccs. The graver has retouch that alternates between dorsal and ventral faces. The combination tool is an end scraper-knife on a dark red ccs cortical flake. Faunal remains from component I (1991 and 1996 excavations) number 18 pieces. Identified taxa include horse/ass (Equus sp.), marmot (Marmota sp.), and large mammal.

Component II

No features were encountered in component II, and, as described above, this component is considered to be redeposited and in a secondary position. The artifact assemblage from this component includes 104 lithic artifacts, 1 ceramic sherd, and 2 small unidentifiable bone fragments. The single ceramic sherd is an undecorated dark gray body shard similar to those described for component III, and is probably intrusive from that overlying stratum. The lithic artifact assemblage is made up of 81 debitage pieces and 23 tools. Raw materials include dark gray ccs (45 percent), speckled gray ccs (29 percent), dark red ccs (16 percent), translucent tan/gray ccs (8 percent), tan ccs (1 percent), and fine-grained gray ccs (1 percent). All but the last two raw materials can be found in local creek and river alluvium.

The debitage part of the assemblage includes 2 cores, 23 cortical flakes, 40 flakes, 8 blades, 1 retouch chip, and 7 splinters. Both of the cores are small bipolar cores manufactured on translucent tan/gray ccs. Among the cortical flakes are 13 primary flakes, 9 secondary flakes, and 1 cortical flake fragment. Even though cores are for the most part absent from the assemblage, the high incidence of cortical flakes (28 percent of the debitage) indicates that primary reduction activities occurred frequently at the Masterov Kliuch site. Among 42 platforms scored, 17 percent are cortical, 76 percent are smooth, and 7 percent are dihedral, further indicating that minimal platform preparation was involved in the manufacture of these artifacts. Among blades, there are three proximal blade fragments, four medial blade fragments, and one complete blade. All of these are made either on dark gray or speckled gray ccs.

Among the 23 retouched artifacts, there are 7 retouched blades, 3 retouched flakes, 3 side scrapers, 2 notches, 2 denticulates, 2 possible burins, 1 cortically backed knife, 1 graver, 1 end scraper, and 1 pointed tool. The retouched blades include one unilaterally and six bilaterally retouched pieces. The three side scrapers include a dejete scraper made on a dark gray ccs blade, a unilaterally retouched side scraper made on a speckled gray ccs cortical flake, and a side scraper fragment on a speckled gray ccs flake. Among the burins is a dark gray ccs blade fragment with a possible laterally burinated edge, as well as a dark gray ccs flake with a possible transversely burinated edge. The single pointed tool is made on a translucent tan/gray ccs blade fragment, and is dorsally retouched along both lateral margins. Retouch invasiveness is relatively low, with 17 (68 percent) of 25 measured tool edges displaying retouch scars that travel less than 4 mm from the tool margin.

Component III

While not a focus of our study, excavations in 1996 uncovered an intact Bronze Age living floor with two preserved features: an unlined hearth and stone-lined pit. The hearth, occurring in square 25S at an elevation of 76-82 cm below datum, consists of a wood charcoal and ash stain in an elongate oval shape, roughly 40-60 cm in diametre and 5 cm thick. A sample of the wood charcoal from this hearth yielded an AMS ^{14}C age estimate of 2895 [+ or -] 45 B.P. (AA-23648). The pit feature is situated in squares 25P and 26P, about 2.5 m east of the hearth feature. This 30-cm-deep pit is shaped like an inverted cone, with the top of the pit measuring about 100 cm in diameter, and the base only 10 cm in diameter. The pit's fill is an organic-rich loam with occasional charcoal flecks, small bone fragments, ceramic sherds, microblades, and several large stones.

The component III assemblage consists of 564 lithic artifacts (10 of which are tools), 23 ceramic sherds, and 12 small unidentifiable bone fragments. Retouched artifacts include four retouched microblades, two retouched blades, two end scrapers, one notch, and one hammerstone. The ceramic sherds are all of the same type, but appear to represent at least two different vessels. These are poorly fired, dark-gray colored ceramics that range from about 4 to 6 mm thick. Decorations include bands of diagonal incisions that consistently measure about 10 mm long. Similar pottery styles have been identified at other sites in the Transbaikal with late Holocene components, including Studenoe, Ust'-Menza, and Altan. Radiocarbon ages on such sites range from about 3500 to 2000 B.P., and are commonly attributed to the Bronze Age.

The Masterov Kliuch site contains two stratigraphically distinct early Upper Palaeolithic components. Component I has been accelerator radiocarbon dated to about 32,500-30,000 B.P., while component II has not been radiocarbon dated and appears to lie in a secondary context. Nonetheless, it too can be tentatively attributed to the early Upper Palaeolithic given technological and typological aspects of its lithic assemblage.

Lithic assemblages are characterised by blade and flake primary reduction technologies, with blade cores being either flat-faced or "end" cores (but not prismatic). Bipolar reduction strategies are also evident. Secondary reduction technologies include unifacial as well as burin techniques. Tool assemblages include retouched blades and flakes, end scrapers, gravers, burins, knives, denticulates, and notches. Component II also yielded a small unifacially worked point on a blade.

The core technologies and tool forms present at Masterov Kliuch are characteristic of the Siberian early Upper Palaeolithic, dated elsewhere to between about 42,000 and 30,000 B.P. at sites like Kara-Bom, Makarovo-4, Malaia Syia, Varvarina Gora, Kamenka, and Tolbaga. These sites in turn represent a widespread complex of flat-faced core and blade industries that spanned inner Asia from Uzbekistan in the west to the Transbaikal and perhaps inner Mongolia in the east during the mid-Upper Pleistocene, perhaps signaling the spread of anatomically modern humans from southwestern Asia.

Masterov Kliuch and Site Formation Processes in Siberia

Nearly all early Upper Palaeolithic sites known from Siberia (e.g., Kara-Bom, Kamenka, Sannyi Mys, Varvarina Gora, Tolbaga) occur in cryoturbated colluvial deposits; thus the geoarchaeological lessons learned at Masterov Kliuch have implications for these sites as well. Through careful excavation, three-point proveniencing of artifacts, conjoining artifacts, and measuring of trend and plunge of large artifacts found in situ, we were able to distinguish different degrees of integrity for the two early Upper Paleolithic components at Masterov Kliuch. Component I is characterised by tight vertical concentration and clustered horizontal distribution of artifacts suggesting a primary context, while component II is characterised by dispersed vertical and horizontal distributions of artifacts suggesting a secondary context.

Further, while we were able to refit some artifacts from component I, no conjoinable artifacts were found in component II. Trend and plunge of artifacts in these components are quite variable, but few artifacts were vertically oriented, suggesting that frost-heaving had not significantly displaced artifacts. Similar geoarchaeological studies are needed at sites like Kara-Bom and Makarovo-4 where artifact concentrations are thought to represent intact early Upper Palaeolithic living floors, and at sites like Tolbaga and Sannyi Mys where rings of stones are interpreted as dwelling features. Our experiences at Masterov Kliuch tell us that the behavioural context of artifacts at these sites, also situated in colluvial sediments along relatively steep slopes, could be disturbed, and that the putative dwelling features could be the product of natural, not cultural, processes. Clearly, reconstructions of early Upper Palaeolithic site structure and settlement behaviour need to proceed with careful consideration of geologic site formation processes.

Masterov Kliuch and Raw Material Procurement in the Siberian Early Upper Palaeolithic

The analysis of the Masterov Kliuch lithic assemblages, although based on a relatively small sample, provides an interesting glimpse into early Upper Palaeolithic raw material selection and procurement. In component I, all lithic artifacts recovered in our excavations are made on raw materials that are available in nearby alluvium of Gyrshelun Creek and the Khilok River, within 2-3 km of the site. Debitage analysis suggests that early Upper Palaeolithic flintknappers carefully selected fine-grained cryptocrystalline-silicate nodules from these sources and carried them to the Masterov Kliuch site for flaking. Core preparation and blade and flake manufacture occurred on the site, as did tool use, resharpening, and discard. There is no evidence of finished tools being transported to the site from some other location, or of exotic raw materials being brought to Masterov Kliuch from more distant sources.

Together, the evidence from component I suggests that early Upper Palaeolithic people were provisioning the Masterov Kliuch site exclusively with local raw materials for the manufacture of stone tools. The site, however, does not appear to have served solely as a task-specific quarry. Instead, evidence for multiple technological activities beyond those expected to be found at a quarry suggests that Masterov Kliuch served more as a residential base than a specialised resource extraction site. Further, the provisioning of this place with only local resources (gathered within 3 km of the site) and the apparent absence of exotic resources indicate that raw material procurement was "embedded" within other foraging activities that were carried out in the immediate area surrounding the site. This pattern of raw material procurement has been noted at other early Upper Palaeolithic sites in Siberia. Most early Upper Palaeolithic sites, including Makarovo-4 in the upper Lena Valley, Arembovskii in the Angara Valley, and Kara-Bom in the Altai Mountains, are situated very near sources of abundant fine-grained cryptocrystalline silicates. In all of these cases, greater than 95 percent of all finished tools are made on local raw materials procured within 5 km of the sites, and the full technological sequence of primary and secondary reduction is represented for these local materials (Goebel 1993, 1999). Further, exotic raw materials are absent from these sites. Like at Masterov Kliuch, then, the lithic assemblages from these early Upper Palaeolithic sites represents embedded raw material procurement strategies that focused on "hyper-local" lithic resources.

Masterov Kliuch and most other Siberian early Upper Palaeolithic sites further appear to represent hunter-gatherer camps that were repeatedly occupied, perhaps because of their proximities to high-quality raw materials as well as because of their ecological settings in areas of high topographic relief and environmental zonation, where diverse animal and plant resources would have been regularly available (Goebel 1999). This pattern of early Upper Palaeolithic raw material procurement and settlement suggests that early modern human hunters of northern Asia were often "tethered" to locations on the landscape where lithic raw materials as well as diverse faunal resources were locally abundant and accessible.

Similar patterns of raw material procurement and settlement have been documented for the Mousterian of southwest Europe (Mellars 1996) and the earliest Upper Palaeolithic complexes of Central Europe (i.e., the Bohunician and Szeletian). In these areas, the transport of exotic raw materials and the more logistical procurement strategies that they represent did not appear until the emergence of the Aurignacian after 35,000 B.P. In Siberia, such behaviours appear to have emerged even later, sometime after 25,000 B.P. during the time of the "Mal'ta Culture," the region's middle Upper Palaeolithic complex (Goebel 1999). Perhaps this means that aspects of logistical organisation and planning, so commonly portrayed to represent modern human behaviour, were relatively late in developing among the Upper Palaeolithic hunter-gatherers of northern Asia.

6

Future Trends in Ocean Sciences

Weather Satellite

The weather satellite is a type of satellite that is primarily used to monitor the weather and climate of the Earth. Satellites can be either polar orbiting, seeing the same swath of the Earth every 12 hours, or geostationary, hovering over the same spot on Earth by orbiting over the equator while moving at the speed of the Earth's rotation. These meteorological satellites, however, more than clouds and cloud systems. City lights, fires, effects of pollution, auroras, sand and dust storms, snow cover, ice mapping, boundaries of ocean currents, energy flows, etc., and other types of environmental information are collected using weather satellites. Weather satellite images helped in monitoring the volcanic ash cloud from Mount St. Helens and activity from other volcanoes such as Mount Etna. Smoke from fires in the western United States such as Colorado and Utah have also been monitored.

Figure: *Weather satellites carry instruments called radiometers*

Other environmental satellites can detect changes in the Earth's vegetation, sea state, ocean colour, and ice fields. For example, the 2002 oil spill off the northwest coast of Spain was watched carefully by the European ENVISAT, which, though not a weather satellite, flies an instrument (ASAR) which can see changes in the sea surface.

El Niño and its effects on weather are monitored daily from satellite images. The Antarctic ozone hole is mapped from weather satellite data. Collectively, weather satellites flown by the U.S., Europe, India, China, Russia, and Japan provide nearly continuous observations for a global weather watch.

Observation

Observation is typically made via different 'channels' of the Electromagnetic spectrum, in particular, the Visible and Infrared portions.

Some of these channels include :

- *Visible and Near Infrared:* 0.6 ìm - 1.6 ìm - For recording cloud cover during the day
- *Infrared:* 3.9 ìm - 7.3 ìm (Water Vapour), 8.7 ìm, - 13.4 ìm (Thermal imaging)

History

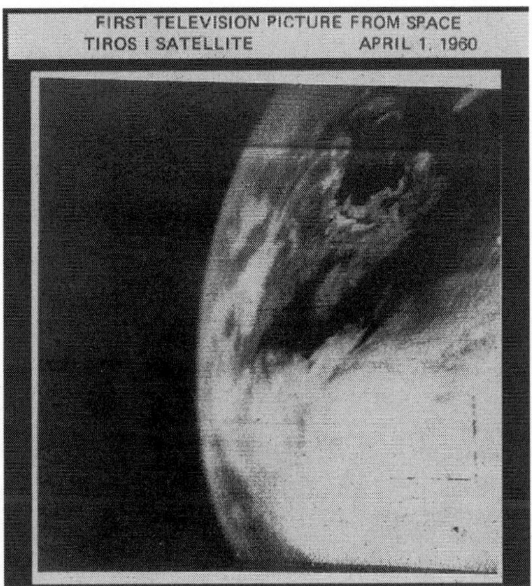

Figure: *The first television image of Earth from space from the TIROS-1 weather satellite.*

The first weather satellite, Vanguard 2, was launched on February 17, 1959. It was designed to measure cloud cover and resistance, but a poor axis of rotation kept it from collecting a notable amount of useful data. The first weather satellite to be considered a success was TIROS-1, launched by NASA on 1 April 1960. TIROS operated for 78 days and proved to be much more successful than Vanguard 2. TIROS paved the way for the Nimbus program, whose technology and findings are the heritage of most of the Earth-observing satellites NASA and NOAA have launched since then.

Visible Spectrum

Visible-light images from weather satellites during local daylight hours are easy to interpret even by the average person; clouds, cloud systems such as fronts and tropical storms, lakes, forests, mountains, snow ice, fires, and pollution such as smoke, smog, dust and haze are readily apparent. Even wind can be determined by cloud patterns, alignments and movement from successive photos.

Infrared Spectrum

The thermal or infrared images recorded by sensors called scanning radiometres enable a trained analyst to determine cloud heights and types, to calculate land and surface water temperatures, and to locate ocean surface features. Infrared satellite imagery can be used effectively for tropical cyclones with a visible eye pattern, using the Dvorak technique, where the difference between the temperature of the warm eye and the surrounding cold cloud tops can be used to determine its intensity (colder cloud tops generally indicate a more intense storm). Infrared pictures depict ocean eddies or vortices and map currents such as the Gulf Stream which are valuable to the shipping industry. Fishermen and farmers are interested in knowing land and water temperatures to protect their crops against frost or increase their catch from the sea. Even El Niño phenomena can be spotted. Using colour-digitized techniques, the gray shaded thermal images can be converted to colour for easier identification of desired information.

Types

There are two basic types of meteorological satellites: geostationary and polar orbiting.

Geostationary

Geostationary weather satellites orbit the Earth above the equator at altitudes of 35,880 km (22,300 miles). Because of this orbit, they remain stationary with respect to the rotating Earth and thus can record or transmit images of the entire hemisphere below continuously with their visible-light and infrared sensors. The news media use the geostationary photos in their daily weather presentation as single images or made into movie loops. These are also available on the city forecast pages of noaa.gov (example Dallas, TX).

Figure: *Image from the GOES-9 weather satellite of Hurricane Felix.*

Several geostationary meteorological spacecraft are in operation. The United States has two in operation; GOES-11 and GOES-12. GOES-12, designated GOES-East, is located over the Amazon River and provides most of the U.S. weather information. GOES-11 is GOES-West over the eastern Pacific Ocean. Russia's new-generation weather satellite Elektro-L 1 operates at 76°E over the Indian Ocean. The Japanese have one in operation; MTSAT-1R over the mid Pacific at 140°E. The Europeans have Meteosat-8 (3.5°W) and Meteosat-9 (0°) over the Atlantic Ocean and have Meteosat-6 (63°E) and Meteosat-7 (57.5°E) over the Indian Ocean. India also operates geostationary satellites called INSAT which carry instruments for meteorological purposes. China operated the Feng-Yunÿ¨˜ò–) geostationary satellites FY-2D at 86.5°E and FY-2E at 123.5°E, which are no longer in use.

Polar Orbiting

Polar orbiting weather satellites circle the Earth at a typical altitude of 850 km (530 miles) in a north to south (or vice versa) path, passing over the poles in their continuous flight. Polar satellites are in sun-synchronous orbits, which means they are able to observe any place on Earth and will view every location twice each day with the same general lighting conditions due to the near-constant local solar time. Polar orbiting weather satellites offer a much better resolution than their geostationary counterparts due their closeness to the Earth.

Figure: *Computer controlled motorized parabolic dish antenna for tracking LEO weather satellites.*

The United States has the NOAA series of polar orbiting meteorological satellites, presently NOAA 17 and NOAA 18 as primary spacecraft, NOAA 15 and NOAA 16 as secondary spacecraft, NOAA 14 in standby, and NOAA 12. Europe has the Metop-A satellite. Russia has the Meteor and RESURS series of satellites. China has FY-1D and FY-3A. India has polar orbiting satellites as well.

DMSP

The United States Department of Defence's Meteorological Satellite (DMSP) can "see" the best of all weather vehicles with its ability to detect objects almost as 'small' as a huge oil tanker. In addition, of all the weather satellites in orbit, only DMSP can "see" at night in the visual. Some of the most spectacular photos have been recorded by the night visual sensor; city lights, volcanoes, fires, lightning, meteors, oil field burn-offs, as well as the Aurora Borealis and Aurora Australis have been captured by this 450-mile-high space vehicle's low moonlight sensor. At the same time, energy monitoring as well as city growth can be accomplished since both major and even minor cities, as well as highway lights, are conspicuous. This informs astronomers of light pollution. The New York City Blackout of 1977 was captured by one of the night orbiter DMSP space vehicles.

In addition to monitoring city lights, these photos are a life saving asset in the detection and monitoring of fires. Not only do the satellites see the fires visually day and night, but the thermal and infrared scanners on board these weather satellites detect potential fire sources below the surface of the Earth where smoldering occurs. Once the fire is detected, the same weather satellites provide vital information about wind that could fan or spread the fires. These same cloud photos from space tell the firefighter when it will rain.

Dramatic photos are provided by all the weather satellites, but even more definitive were the DMSP night visible-light pictures of the 700 oil well fires that Iraq started on 23 February 1991 as they fled Kuwait. These fires were vividly illustrated as huge flashes in the night photos, far outstripping the glow of large populated areas. The fires consumed millions of gallons of oil; the last was doused on November 6.

Uses

Snowfield monitoring, especially in the Sierra Nevada, can be helpful to the hydrologist keeping track of how much snow is available for runoff vital to the water sheds of the western United States. This information is gleaned from existing satellites of all agencies of the U.S. government (in addition to local, on-the-ground measurements). Ice floes, packs and bergs can also be located and tracked from weather space craft.

Even pollution whether it's nature-made or man-made can be pinpointed. The visual and infrared photos show effects of pollution from their respective areas over the entire earth. Aircraft and rocket pollution, as well as condensation trails, can also be spotted. The ocean current and low level wind information gleaned from the space photos can help predict oceanic oil spill coverage and movement. Almost every summer, sand and dust from the Sahara Desert in Africa drifts across the equatorial regions of the Atlantic Ocean. GOES-EAST photos enable meteorologists to observe, track and forecast this sand cloud. In addition to reducing visibilities and causing respiratory problems, sand clouds suppress hurricane formation by modifying the solar radiation balance of the tropics. Other dust storms in Asia and mainland China are common and easy to spot and monitor, with recent examples of dust moving across the Pacific ocean and reaching North America. In remote areas of the world with few local observers, fires could rage out of control for days or even weeks and consume millions of acres before authorities are alerted. Weather satellites can be a tremendous asset in such situations. Nighttime photos also show the burn-off in gas and oil fields.

High Resolution Sea Surface Temperature Analysis for the Arctic Ocean

The Sea surface temperature (SST) is a vital parameter that is used in both oceanic models and Numerical Weather Prediction systems. The temperature of the sea surface can be observed from space using both infrared and microwave sensors. The infrared sensors typically have the best spatial resolution and highest accuracy but they are limited by cloud cover. In the microwave part of the spectrum, the SST can be observed in the presence of cloud cover, but the spatial resolution is not as good as for the infrared satellite sensors. The different sampling characteristics of the satellites and the model demand for high resolution SST fields without gaps, makes it very relevant to perform an analysis of the SST observations, whereby the different satellite observations are referenced to each other and interpolated to produce a daily high resolution field without gaps. For this purpose, DMI has developed an optimal interpolation algorithm that uses statistics derived from the data itself to produce level 4 fields for the North Sea Baltic Sea, the Arctic Ocean and several other regions around the world. The analysis system for the Arctic Ocean is used in this paper as an example but the other analysis products are very similar except for the region specific statistics.

The Dmi Analysis System

The DMI analysis system consists of several different steps. The major tasks are:

1. Retrieval of the satellite L2P observations
2. Pre-processing: quality control, collating the satellite observations
3. Optimal Interpolation, including ice masking
4. Generation of monitoring and validation statistics.

A fundamental part of the scheme is the ingestion of the satellite data. All the satellite observations that are used to produce the SST fields are obtained via ftp from the GHRSST-pp project as level 2P observations (Donlon et. al, 2007).

Table 1: *Satellite observations that are currently included in the DMI level 4 analysis system for the Arctic Ocean. IR stands for infrared and MW stands for Microwave.*

Sensor	*Satellite*	*Sensor*	*Resolution*
AATSR	ENVISAT	IR	1 km
MODIS	Aqua	IR	1km
Modis	Terra	IR	1km
AVHRR	NOAA 17+18	IR	2km + 9 km
Seviri	MSG-1	IR	5 km
AMSR-E	Aqua	MW	25 km

In addition to the satellite SST observations, ice information is included from the Ocean and Sea Ice SAF project. The ice product is the Northern Hemisphere ice edge product in a 0.1 degrees resolution.

The individual L2P data products are being quality controlled, corrected for biases and averaged to the analysis grid before they are being used in the analysis. All satellite observations within 24 hours from the analysis time is included in the Arctic analysis. Both day and night time observations are included and all observations are referenced to subskin observations.

In the analysis step, the collated files are ingested and anomalies are generated by subtracting the previous days analysis as a guess field. All grid points with ice present are treated as satellite observations with an SST of -1.8°C. If more that one satellite observation is available in a grid point, a weighted average is calculated for the given point using the individual error values. Optimal interpolation is performed on the resulting collated grid to produce the SST analysis and error estimates. After the interpolation, monitoring statistics is generated and figures produced about e.g. the number of satellite observations included, anomalies of the individual satellite products, the size of the increments etc.. Finally, a validation routine compares the drifting buoy observations to the SST analysis to produce error statistics.

The analysis system has been running for the Arctic Ocean in a 0.05 degrees resolution. Analysis fields are available from January 2006 up to present.

Validation Against in Situ Observations

Validation of the SST products is included in the operational system. The positions and anomalies of the drifting buoys for the first half of the year in 2008.

It is clear from the figure that the performance of the Arctic Level 4 SST product is good in terms of standard deviation, whereas the bias seems to be too high in the beginning of the year. However the quality control on the buoy observations is not adequate, as there are some matchups that are situated in regions with permanent ice cover. These positions probably correspond to buoys located on drifting ice, measuring the air temperature. The validation should therefore be interpreted as a high estimate on the error on the SST product and more rigorous quality control on the in situ observations will be part of the future work.

Marginal Ice Zones

There are several factors in the Arctic Ocean that makes the production of an accurate L4 SST product difficult. Among these are:

- Very few SST retrievals in the vicinity of the ice edge.
- Altered error covariance in the marginal ice zone
- Questionable accuracy of the SST observations in the marginal ice zone.
- Large gradients in SST in the vicinity of the ice edge (e.g. in the East Greenland Current).

The future work with the Arctic Ocean will focus upon resolving and determining these issues in the marginal ice zone. The DMI processing system carries information about the age of the satellite observations in each grid point. As an example of the poor retrieval. All data within the marginal ice zone are at least 2 days old, outlining the very poor data return in the vicinity of the ice edge.

The DMI analysis system ingests all available GHRSST-pp L2P products to produce daily high resolution SST products for the Arctic Ocean. The validation of the SST product against drifting buoys shows a standard deviation of around 0.6°C and a positive bias, which is highest in the wintertime and decreasing in the spring time. The error estimates are high limits and a more careful selection of the buoy observations should discard obviously erroneous observations. In the Arctic Ocean, there are several factors that complicate the production of an accurate SST analysis product, such as the quality of the marginal ice zone SST retrievals, and altered statistics close to the ice edge. These issues will be included in the future work in order increase the quality of the DMI Arctic Ocean SST analysis.

Skin and Bulk Sea Surface Temperature Estimates from Passive Microwave and Thermal Infrared Satellite Imagery

The current generation of infrared and passive microwave satellite sensors provide highly complementary information for monitoring sea surface temperature (SST). On the one hand, infrared sensors provide high resolution and high accuracy but are obscured by clouds. Microwave sensors on the other hand, provide coverage through non-precipitating clouds but have coarser resolution and generally poorer accuracy. The primary sources of retrieval uncertainty are also different for the two sensors. These factors have motivated recent work in blending the data in an attempt to produce improved SST products. Prior to combining the data, however, the source of the differences between SST retrievals from the two sensors must be fully understood. This is especially significant since the sensors effectively respond to the temperature at slightly different depths in the water column. The main focus of this study is to examine the exact character of the differences between representative infrared and passive microwave SST products, and their dependencies on atmospheric forcing such as wind speed, water vapour, aerosols, and atmospheric stability.

Data

A detailed comparison is performed between existing infrared and microwave sea surface temperature products and independent in situ observations from drifting and moored buoys. The infrared product is the operational AVHRR nonlinear SST (NLSST) product from the Naval Oceanographic Office. The passive microwave product is the TRMM Microwave Imager (TMI) SST retrievals prepared by Remote Sensing Systems. The buoy observations were drawn from an archive of NCEP GTS surface marine data maintained at the NOAA Climate Diagnostics Centre. Extensive additional quality control was applied to the buoy data to eliminate erroneous observations and transmission errors. Two years of data from 1999-2000 were used in the comparisons.

Differences were computed between the two satellite products and between each satellite product and the buoy observations. For the comparisons with buoys, sets of matches were constructed where the satellite and surface observations were collocated within 25 km and 1.5 hours. For the satellite intercomparisons, separate daytime and nighttime 0.25 resolution grids from each day were differenced. The corresponding difference grids were then averaged over monthly periods.

SST Product Differences and Dependencies

Initial overall evaluation of the AVHRR and TMI SST products with buoy observations suggest that both products perform well. For both products, overall biases and RMS differences are relatively small.

Table : *Overall Statistics for TMI and AVHRR SST Products*

SST Product	*Bias (deg C)*	*RMS (deg C)*
TMI SST	0.12	0.71
AVHRR SST	0.03	0.54

Maps of the differences between the products, however, show complex spatial and temporal differences. Average nighttime differences between the products for June 2000. Note the strong spatial correlation of the differences and the changes between seasons. Clearly such differences are not random and need to be understood and accounted for if the products are to be combined.

The differences can result both from retrieval errors and actual geophysical processes. To begin to account for the differences, potential sources of error in the retrievals were explored in detail. The matchups constructed between the satellite retrievals and buoy observations were used to explore the dependence of errors in the retrievals with multiple environmental parameters. Parameters found to impact the retrievals included wind speed, water vapour, SST, aerosols, and atmospheric stability. These results were then used to derive bias adjustments and uncertainty estimates as functions of the environmental parameters.

A primary source of uncertainty in microwave SST retrievals is variations in surface roughness and emissivity resulting from wind speed changes.

The microwave retrieval approach uses multiple frequencies, some of which are sensitive to variations in total column water vapour content. As such, it is conceivable that water vapour variations could also influence the microwave SST retrievals. Similarly, since sea surface emissivity varies with temperature, retrieval errors could also exhibit dependence on SST.

Since water vapour and SST are not independent, and to facilitate application of bias adjustments to the TMI retrievals, the TMI SST and buoy differences were binned into small ranges of wind speed, water vapour, and SST, and three-dimensional look-up tables were derived for the bias and RMS difference as a simultaneous function of all parameters. Bias adjustments to the TMI retrievals based on the three-dimensional look-up table are denoted by. While there is no apparent wind speed dependence observed for the wind speed range between 4 and 12 m/s, slight positive increases in bias and scatter are observed at high wind extremes, and a negative bias relative to buoys is observed at the low wind extremes. The buoy comparisons also suggest some dependence of the TMI retrievals on both SST and water vapour. In particular, large positive biases are observed at the warmest SSTs. These results were additionally screened for proximity to land but increasing the allowable separation did not significantly alter the results.

The primary function of infrared SST retrieval algorithms is to correct for the effects of atmospheric water vapour and, therefore, water vapour could remain a source of error in the algorithms. Because bias adjustments based on water vapour alone did not appear to capture all variability in the differences relative to buoys and because of the high correlation between water vapour and SST, the AVHRR - buoy differences were evaluated as a simultaneous function of water vapour and SST. Overall biases are small but the results suggest some residual dependence on water vapour and SST.

Interestingly, the shape of the dependence is similar to that for the microwave retrievals. The dependence on vapour is greatest at low water vapour while SST has most impact when high. Since a distinct retrieval algorithm is used during the daytime, a separate daytime dependence was also determined. As before, a two-dimensional look-up table was derived to facilitate the bias adjustments to the AVHRR SST retrievals.

Infrared retrievals are sensitive to aerosols whereas microwave ones are not. As a result, comparisons between the infrared SSTs and the independent microwave measurements were performed to see if a relationship with marine aerosols could be identified. An estimate of aerosol optical depth (aod) over the oceans derived from the NOAA/NESDIS AVHRR aerosol operational product was used in these comparisons. The differential bias was applied as another correction in the retrieval of a more accurate estimate of the infrared SSTs.

Application of bias adjustments to the SST products based on these dependencies still left significant structure in the remaining differences particularly in the cool-water, upwelling regions off the west coasts of North America, South America, and Africa.

The locally cool water provides the potential for greater atmospheric stability than in the surrounding regions. Atmospheric stability can potentially influence microwave retrievals since increased stability can damp ocean surface roughness. Air temperatures derived from the Oregon State University Climate Research Institute (OSUSFC) Climatology were used in conjunction with the TMI SSTs to obtain an estimate of sea-air temperature difference (dtsa). The identified bias dependence was applied as another correction to the TMI SST retrievals.

Impact of Product Dependencies

Using bias adjustments derived from each of the observed product dependencies, revised difference maps between the TMI and AVHRR SST products were computed. The results corresponding to June 2000. The overall differences after the bias adjustments are applied are smaller and less spatial coherence is observed relative to the direct comparison. The lowest RMS differences are observed after all corrections have been applied. Despite the improvement, differences between the products can still be observed. The remaining differences, mostly in the RMS, could be caused by geophysical differences related to the different effective measurement depths of the two sensors. Additional work is ongoing to explore this possibility.

Ocean Surface Topography

Orbiting spacecrafts make highly accurate measurements of the height of the ocean surface - commonly called ‘sea level’ - to gather long-term information about the world’s ocean and its currents. These measurements provide information about the topography of the ocean’s surface, which is used to study weather, climate, and other dynamic ocean phenomena. Ocean surface topography data also have many other applications, such as in fisheries management, navigation and offshore operations.

Ocean and land topography are defined exactly the same way. Both give the height of the ocean or land above the geoid. The geoid is the shape the sea surface would have if all the currents and tides stopped.

Accurate measurements of ocean surface topography are important for studying ocean tides, circulation and the amount of heat the ocean holds. The observations are used to help predict short-term changes in weather and longer-term climate patterns.

Tides and currents are the two major factors that contribute to the dynamic topography of the ocean's surface. If our oceans had no tides or currents, the sea surface would assume the shape of the geoid.

- Tides result from changes in gravity at Earth's surface due to the moon and Sun. Detailed information about tides can be found at NOAA's Centre for Operational Oceanographic Products and Services web site.
- Ocean currents are streams of water flowing through the ocean, driven by wind or by the mixing of waters of differing densities. Because currents are moving bodies of water, they result in changes in the ocean surface, altering ocean surface topography by a few tens of centimetres to more than a metre. Although on a global scale this seems like a small amount, NASA's ocean altimetre satellites can easily measure these changes, allowing data from the satellites to be used to study global ocean circulation, climate and other important areas of research.

NASA currently has two satellite missions that measure ocean surface topography. Jason-1, launched in 2001 continues the measurements begun by TOPEX/Poseidon, which operated from 1992 through 2006. The follow on mission to Jason-1, the Ocean Surface Topography Mission on Jason-2 (OSTM/Jason-2), was launched in 2008 and will take this important data record into a second decade.

These missions use radar altimetre systems specially designed to make extremely accurate and precise measurements of the height of the ocean surface. The satellites measure the height of the sea surface with an accuracy of about 3 centimetres (just over 1 inch) relative to the centre of the Earth, and they collect data over nearly all of Earth's ice-free ocean every 10 days. This illustration of an altimetry satellite shows its orbit and its relationship to the sea surface, geoid and seafloor. The second component of the ocean height measurement is the range from the satellite to the ocean surface. To take a measurement, an onboard altimetre bounces microwave pulses off the ocean surface and measures the time it takes the pulses to return to the spacecraft. The highly accurate altimetre range measurements are subtracted from the satellite orbital height, resulting in ocean topography measurements that are accurate to 3.3 centimetres (1.3 inches) relative to the centre of the Earth. By averaging the few-hundred thousand measurements collected by the satellite in the time it takes to cover the global ocean (10 days), global mean sea level can be determined with a precision of several millimetres.

For as long as both the Jason-1 and OSTM/Jason-2 satellites remain operational, they will orbit Earth in a tandem formation separated by about 330 kilometres, doubling the science data and bringing us closer to solving the global climate puzzle.

NASA and the French space agency, CNES, are joint partners in the Jason-1 mission. OSTM/Jason-2 is a multi-partner collaboration among NASA, CNES, NOAA and the European Organisation for the Exploitation of Meteorological Satellites (EUMETSAT) that will continue the long-term ocean surface topography data record. NASA altimetry data are often combined with data from other ocean-viewing satellites and from in situ measurements taken from ships, profiling buoys and drifters. These combined data sets are used to:

- Measure global sea-level change and provide a continuous view of changing global ocean surface topography
- Calculate the transport of heat, water mass, nutrients and salt by the oceans
- Increase understanding of ocean circulation and seasonal changes and how the general ocean circulation changes through time
- Provide estimates of significant wave height and wind speeds over the ocean
- Test how we compute ocean circulation caused by winds
- Improve the knowledge of ocean tides and develop global tide models
- Improve forecasting of climatic events like El Niño and La Niña as well as global climate change
- Describe the nature of ocean dynamics and develop a global view of Earth's ocean
- Monitor the variation of global mean sea level and its relation to global climate change

Ocean surface topography data have a wide spectrum of scientific applications. Altimetre data have been instrumental in observing the rise and fall of the El Niño and La Niña and, more recently, the Pacific

Decadal Oscillation. Ocean topography often corresponds to the amount of heat stored in the upper layers of the ocean. This information is useful in predicting hurricane season severity and forecasting individual storm severity. Altimetry data also provide crititical contributions to our knowledge of even longer-term phenomena such as global sea level rise. Over the next few decades, scientists will continue to study these climate, weather and ocean circulation phenomena to better understand their impact on our environment and our daily lives.

In addition to their value as a research tool, ocean surface topography data also have many practical and operational applications. They are used in ship routing, offshore oil operations, fisheries management, recreational boating, oil spill remediation, as well as search-and-rescue and naval maritime operations. Ocean altimetre data have been used to help understand the behaviour of Stellar sea lions and the migratory habits of hawksbill turtles and whales tracked by other satellites. The information ocean topography data provide on the extent and degree of ocean warming is important for coral reef research. There are also land applications of ocean altimetry data. Near real-time data of the height of water in rivers and lakes is used for flood and drought investigations. Altimetry also provides a measurement of river level variations, which is especially useful in areas that are difficult to reach such as the Amazon River basin. Altimetry data also help to monitor Greenland ice sheet growth rates.

Future

Scientists and engineers working on the OSTM/Jason-2 mission have developed technologies and data methodologies that will result in even more accurate measurements of sea-surface height than previous missions could provide. The new altimetre systems can observe changes in ocean surface topography with an accuracy of one to two centimetres. Ocean surface topography data are also being combined with other measurements to help scientists and forecasters better understand the link between the ocean and short and long-term changes in climate and how Earth's climate may respond to changes in carbon dioxide in the atmosphere and to man-made changes on land.

Temperature

Instruments aboard NASA's and NOAA's spacecrafts use their vantage point from space to collect global measurements of the ocean's surface temperature. Each day these instruments make thousands of measurements of broad swaths of the Earth - creating concurrent data sets of the entire planet. By developing global, detailed, and decades-long views of Sea Surface Temperature (SST), data obtained from NASA and NOAA satellites provide the basis for the prediction of climate change, ocean currents, and the potent El Niño-La Niña cycles.

El Niño is perhaps the best known example of the impact that changing sea surface temperature has on our climate. Every three to seven years, this warming of surface ocean waters in the eastern tropical Pacific brings winter droughts and deadly forest fires in Central America, Indonesia, Australia, and southeastern Africa, and lashing rainstorms in Ecuador and Peru. El Niño affects thousands of people worldwide, and billions of dollars in economic impact. El Niño's "sister," La Niña, occurs less frequently and has the opposite effect - the cooling of surface ocean waters.

But changing SST patterns have broader implications than just the El Niño and La Niña cycles. Changes in SST are the single most important indicator of climate change. Heat is one of the main drivers of global climate, and the ocean is a huge reservoir of heat. The top 6.5 feet of ocean has the potential to store the equivalent amount of heat contained in the atmosphere. The ocean has a high capacity and as ocean currents move tremendous amounts of water over vast distances, heat is also carried or transferred over these distances. This release of heat can play a major role in climate from the regional/basin to global scale. It is for this reason that oceans are termed the 'memory' of the Earth's climate system. Tracking SST as a variable over long periods of time, as well as operationally, is critical for developing climate models and improved weather forecasts.

These cold, deep waters upwell along an equatorial swath around and to the west of the Galapagos Islands. Also noticeable are warm, wide currents of the Gulf Stream moving up the United States' east coast, carrying Caribbean warmth toward Newfoundland and across the Atlantic toward Western Europe. Additionally, there is a warm tongue of water extending from Africa's east coast to well south of the Cape of Good Hope. The distribution of temperature at the sea surface tends to be zonal, that is, it is independent of longitude. Uneven heating of the Earth by the Sun causes the warmest water to be near the equator, while the coldest water is near the poles. The deviations from these zonal measurements are small. The anomalies of sea-surface temperature, the deviation from a long term average, are also small, less than 1.5°C/34.7°F except in the equatorial Pacific where the deviations can be 3°C/37.4°F. Large deviations in the Equatorial Pacific are due primarily to the El Niño-La Niña cycle.

Most weather and climate events are the result of sea and atmospheric coupling. Heat energy released from the ocean is the dominant driver of atmospheric circulation and weather patterns. SST influences the rate of energy transfer into the atmosphere, as evaporation increases rapidly with temperature. Knowing the temperature of the ocean surface provides tremendous insight into short and long term weather and climate events.

Taking the Ocean's Temperature

The most commonly used instrument to measure sea-surface temperature from space is the Advanced Very High Resolution Radiometre AVHRR. Since 1999, the Moderate-resolution Imaging Spectroradiometre (MODIS) sensor has been collecting even more detailed measurements of surface temperature. More recently, the Advanced Microwave Scanning Radiometre for EOS (AMSR-E) has been collecting SST that includes areas covered by clouds. A key attribute of the AVHRR data is the length of the time record. An AVHRR sensor has been carried on all polar-orbiting meteorological satellites operated by NOAA since Tiros-N was launched in 1978. High quality measurements of the temperature of the ocean are now available from 1981 to the present. This unique SST data set is now the longest satellite derived oceanographic record, providing a 25-year (and continuing) record of global SST changes. Conversely, MODIS and records are much shorter.

Thermal Infrared Remote Sensing

AVHRR and MODIS instruments use radiometres to measure the amount of thermal infrared radiation given off by the surface of the ocean. Thermal infrared remote sensing is based on the fact that everything above absolute zero (-273°C/459°F) emits radiation in the thermal infrared region of the electromagnetic spectrum. The amount of thermal infrared radiation given off by an object is related to its temperature (dying embers give off less radiation than a hot fire). Thus by measuring the amount of radiation given off by the ocean we can calculate its temperature. With instruments like radiometres, it is possible to get a picture of the thermal environment that we cannot experience with our normal human sensors. The ability to record precise variations in infrared radiation has tremendous application in extending our observation of many types of phenomena where minor temperature variations are significant in understanding our environment.

The warmer temperatures of 30 - 35 degrees Celsius and are located primarily in the equatorial regions with the greatest concentration in the Indian Ocean and western Pacific. Temperatures near 0 degrees Celsius are shown as black and are concentrated near the polar regions. Sea surface temperature data are used to help us predict weather patterns, to track ocean currents, and to monitor El Niño and La Niña. Sea surface temperature influences the growth of phytoplankton, as well as precipitation patterns across continents, thus indirectly influencing land vegetation. (Data from Advanced Very High Resolution Radiometre [AVHRR]).

Moderate-resolution Imaging Spectroradiometre (MODIS)

MODIS is sensitive to five different wavelengths, or "channels," of radiation used for measuring SST. Both night and day, the sensor measures the thermal infrared energy escaping the atmosphere at 12 microns and then compares that measurement to how much energy is escaping at 11 microns, allowing scientists to determine how much the atmosphere modifies the signal so they can "correct" the data to more accurately derive SST. The MODIS sensor, because of the increased number of channels, tells us a great deal about the influence of the atmosphere on measurements of SST. Similar to AVHRR, MODIS also takes daily measurement of the global ocean.

Advanced Microwave Scanning Radiometre for EOS (AMSR-E)

Because AVHRR and MODIS cannot observe the ocean when the atmosphere is cloudy, NASA developed a new sensor, AMSR-E, that is able to observe through the clouds. AMSR-E on the Aqua satellite is a passive microwave radiometre, modified from the Advanced Earth Observing Satellite-II (ADEOS-II). Microwaves are radio waves that are able to pass through clouds. Thus, the AMSR-E instrument can measure radiation from the ocean surface through most types of cloud cover, supplementing infrared based measurements of SST that are restricted to cloud-free areas. However, the resolution of AMSR-E is coarser than the thermal IR sensors. The addition of AMSR-E data will provide a significant improvement in our ability to monitor SST and temperature controlling phenomenon.

Sea Surface Height & Temperature

Sea surface height data can also provide clues to studying the temperature of the ocean. Warm water expands raising the sea surface height. Conversely, cold water contracts lowering the height of the sea surface. Thus, measurements of sea surface height can provide information about the heat content of the ocean. The height can tell us how much heat is stored in the ocean water column below its surface. Learn more about sea surface height.

The temperature of the world's ocean surface provides a clear indication of the regions where hurricanes and typhoons form, since they can only form when the sea surface temperature exceeds 27.8°C (82°F). In this visualisation of AMSR-E data covering the period from June, 2002 to September, 2003, areas with surface temperatures greater than 82°F are shown in yellow and orange, while sea surface temperatures below 82°F are shown in blue. The region in the Atlantic from the Caribbean to the equator only exceeds the critical temperature during late summer and early fall in the Northern Hemisphere, the period known as "Hurricane Season." It is also possible to see the Gulf Stream, the warm river of water that parallels the east coast of the United States before heading towards northern Europe, in this data. Around January 1, 2003, a cooler-than- normal region of the ocean appears just to the west of Peru as part of an La Niña and flows westward, driven by the trade winds. The waves that appear on the edges of this cooler area are called tropical instability waves and can also be seen in the equatorial Atlantic Ocean at about the same time. This animation shows how the sea surface temperature can cause hurricanes to form. Areas shown in orange and yellow are above 82 degrees F (27.8 degrees C) which is required for hurricanes to be able to form. Sea surface temperatures below 82 degrees F are shown in blue. The data for this visualisation is from the AMSR-E instrument from June 2002 to September 2003.

Interpreting Sea Surface Temperature Measurements

Radiation observed by AVHRR and MODIS is modified by its passage through the atmosphere. The degree to which the signal is modified depends upon the chemistry of the overlying atmosphere. Clouds, haze, dust or smoke can interfere with a space-based remote sensor's ability to accurately measure SST, as can greenhouse gases, like water vapour. These are present in abundance in the tropics and strongly absorb infrared energy and re-radiate it back toward the surface. Scientists have created several algorithms to correct the impact of these variables creating more accurate measurements of SST.

Further, scientists analyse SST data to provide new products that have a wide variety of uses. SST data are also distributed and processed by several organisations. These data sets are then used operationally by sponsoring agency scientists and other organisations.

- The Goddard Earth Sciences Data and Information Services Centre (GES DISC) at CaltTech/JPL are the key distribution point for SST data and related data sets from NASA.
- The Goddard Distributed Active Archive Centre (GDAAC) is the primary distribution centre for MODIS data.
- The Global Hydrology & Climate Centre provides browse images and some Level 2 AMSR-E data products
- The GES DISC is the mirror site for the level 3 data sets.

SST data is also combined with other data taken in-situ by ships and bouys. This data helps calibrate the satellite data to create a more accurate measurement of SST.

SST data is used by many different organisations for regional studies, anomaly studies, climate and meteorological studies, and to provide near real - time access to the data. SST data products are also widely used by the fishing industry to track the conditions where fish are most likely to be found.

Long term averages of sea surface temperature are used to calculate the normal seas surface temperature conditions for a specific time of year and location. Deviations from the long-term mean are called anomalies. The long-term means are also used for studying climate change. Other data is made available in time intervals of less than a day - in some instances within a few hours of collection. This type of data is mostly used for detecting specific features in the ocean, such as currents and eddies.

The Gulf Stream, one of the ocean's most significant and fastest currents moves at four miles per hour. This current of warm water, which is called the North Atlantic Drift after it turns offshore at Cape Hatteras, travels from the Gulf of Mexico to northern European waters. The current's warm waters are responsible for the more temperate climates experienced by Ireland, England, Scotland, and the Isles of Scily. At the beginning of the Current, in the Gulf of

Mexico, its temperature is around 27°C (80°F), but by the time it reaches northern Europe it has cooled down to a few degrees Fahrenheit. As it cools, it gives up heat to the atmosphere, which carries the warmth to Europe. This false-colour Terra/MODIS image produced from Direct Broadcast data by the Space Science and Engineering Centre at the University of Wisconsin-Madison is made from data on Band 31. Its bright colours represent temperature data of the water: the Gulf Stream is shown in shades of orange and yellow, which puts it near 20°C (68°F) on the scale provided in the image. Land temperature data are not shown in this image and appear black, while white represents areas where no data were collected (because clouds obscured the water).

Trends we Observe

SST data is used to observe many regional phenomena around the world, including the Chesapeake Bay and the Gulf of Mexico, the Gulf Stream, Kuroshio, the Somali Current, the Brazil Current and the East Australian Current. These currents are associated with sharp changes in SST which can be detected using satellites. Coastal water studies are made off the Hawaiian and Alaskan coasts. Multiple studies are also conducted off the North Atlantic.

Sea surface temperatures in the equatorial Pacific affect precipitation (and therefore plant growth) over much of the North American continent. Warmer-than-normal water in the central and western equatorial Pacific, creates higher precipitation in southern and central North America. Conversely, cold water temperatures in the Pacific lead to a decrease in precipitation over northern North America.

SST may also affect one of the world's key large-scale atmospheric circulations - the circulation that regulates the intensity and breaking of rainfall associated with the South Asian and Australian monsoons.

Projects are underway that combine data from multiple satellite systems to produce a robust set of sea surface data for assimilation into ocean forecasting models of the waters around Europe and also the entire Atlantic Ocean. The Global Ocean Data Assimilation Experiment, GODAE, is assimilating sea-surface temperature data, altimetre data, scatterometre data, and drifter data into coupled ocean/atmosphere numerical models to produce forecasts of ocean currents and temperatures up to 30 days in advance everywhere in the ocean.

Future

Finally, projects are also being conducted to combine SST data from various sensors to create the highest quality SST. These projects will create a new generation of multi-sensor, high-resolution SST products. An example of such a project is the GODAE High Resolution Sea Surface Temperature Pilot Project (GHRSST-PP).

SST data are important to the development and testing of a new generation of computer models in which the interacting processes of the land, the atmosphere, and the oceans are coupled. The measurements are widely used in the creation of more accurate weather forecasts and increasingly it is seen as a key indicator of climate change. It is anticipated that projects like GHRSST will provide even higher quality data sets for such things as hurricane forecasting.

Projects like GHRSST lay the groundwork for future cooperations, between NASA and NOAA, as well as internationally. Such cooperations will lead to major innovations in how data is distributed in near real-time, searched and stored. Plans include a joint NASA/NOAA effort to provide users with an interface for accessing both near real-time and historical data for climate studies. Future technologies should allow managers, decision makers, and modellers to search and access data in near real time for specified areas of interest. Additionally, the merging of SST data from different sensors will provide high resolution SST data suitable for coastal studies and management.

Salinity

Although everyone knows that seawater is salty, few know that even small variations in Sea Surface Salinity (SSS) can have dramatic effects on the water cycle and ocean circulation. SSS tells us the about the concentration of dissolved salts in the upper centimetre of the ocean surface. Throughout Earth's history, certain processes have served to make the ocean salty. The weathering of rocks delivers minerals, including salt, into the ocean. Evaporation of ocean water and formation of sea ice both increase the salinity of the ocean. However these "salinity raising" factors are continually counterbalanced by processes that decrease salinity such as the continuous input of fresh water from rivers, precipitation of rain and snow, and melting of ice.

Illustration of the global distribution of sea surface salinity. Highest concentrations (over 37 practical salinity units) of salt water are present the mid-Atlantic Ocean and lower-Atlantic off the coast of Brazil, the Mediterranean Sea and the Red Sea. Lower concentrations are found near the Arctic and Antarctic and the coastal regions of east Asia and western North America. Surface salinities of the ocean. The distributions of salinity are quite different from temperature. High concentrations are usually in the centre of the ocean basins away from the mouths of rivers, which input fresh water. High concentrations are also in subtropical regions due to high rates of evaporation (clear skies, little rain, and prevailing winds) and in landlocked seas in arid regions. At high latitudes, salinity is low. This can be attributed to lower elevation rates and the melting of ice that dilutes seawater. To sum up, salinity is low where precipitation is greater than evaporation, mainly in coastal or equatorial regions. Credit: NASA, Jet Propulsion Laboratory.

Salinity & the Water Cycle

Understanding why the sea is salty begins with knowing how water cycles among the ocean's physical states: liquid, vapour, and ice. As a liquid, water dissolves rocks and sediments and reacts with emissions from volcanoes and hydrothermal vents. This creates a complex solution of mineral salts in our ocean basins. Conversely, in other states of ocean water such as vapour and ice, water and salt are incompatible: water vapour and ice are essentially salt free.

Since 86% of global evaporation and 78% of global precipitation occur over the ocean, SSS is the key variable for understanding how fresh water input and output affects ocean dynamics. By tracking SSS we can directly monitor variations in the water cycle: land runoff, sea ice freezing and melting, and evaporation and precipitation over the oceans.

Yet, at present, our knowledge of salinity on a global scale is extremely limited because we cannot measure salinity from a satellite and are currently restricted to measuring salinity at the planet's surface via ships and buoys.

Salinity & Ocean Circulation

Changes in salt concentration at the ocean surface affect the weight of surface waters. Fresh water is light and floats on the surface, while salty water is heavy and sinks. Together, salinity and temperature determine seawater density and buoyancy, driving the extent of ocean stratification, mixing, and water mass formation.

Greater salinity, like colder temperatures, results in an increase in ocean density with a corresponding depression of the sea surface height. In warmer, fresher waters, the density is lower resulting in an elevation of the sea surface. These height differences are related to the circulation of the ocean. The changes in density bring warm water poleward on the surface to replace the sinking water driving the global thermohaline (heat & salt) circulation within the ocean called the Global Conveyor Belt.

Generalised model of the thermohaline circulation: 'Global Conveyor Belt' This illustration shows cold deep high salinity currents circulating from the north Atlantic Ocean to the southern Atlantic Ocean and east to the Indian Ocean. Deep water returns to the surface in the Indian and Pacific Oceans through the process of upwelling. The warm shallow current then returns west past the Indian Ocean, round South Africa and up to the North Atlantic where the water becomes saltier and colder and sinks starting the process all over again.

"The Global Conveyer Belt for Heat" represents in a simple way how ocean currents carry warm surface waters from the equator toward the poles and moderate global climate. This global circuit takes up to 1,000 years to complete.

Salinity & Climate

The Global Conveyor Belt is the principal mechanism by which the oceans store and transport heat. The ocean stores more heat in the uppermost 3 metres than that of the entire atmosphere and acts as a "global heat engine." Since salinity is a key ingredient in the global thermohaline circulation, SSS will help us discover how climate variation induces change in global ocean circulation.

Measuring Salinity

Despite all the progress in understanding our ocean-atmosphere system, Sea Surface Salinity (SSS) - the principal surface tracer of fresh water input and output from the ocean and a direct contributor to seawater density — is not currently measured remotely from space.

A global understanding of SSS has been difficult because sampling by ships, buoys, drifters, and moorings has been extremely limited. Between 300 and 600 AD, awareness of changes in salinity, temperature, and smell helped Polynesians explore the southern Pacific Ocean. In the 1870s, scientists aboard H.M.S. Challenger systematically measured salinity, temperature, and water density in the world's oceans. Over the years, techniques for measuring such ocean water properties have changed drastically in method and accuracy.

Eight days & 100 years of Salinity Data. Salinity has been sparsely detected at sea, limited mostly to summertime observations in shipping lanes. Looking at a map of Earth with 100 years of sea surface data, about 25% of ice-free oceans have never been sampled. During its first two months in space, the Aquarius mission will acquire as many SSS measurements as had been collected from ships and buoys during the previous 125 years. During its three-year mission, it will provide maps of seasonal and year-to-year variation in global SSS; this pioneering information will be used to discern longer-term changes in our oceans and climate. The entire ice free ocean is measured. Since data collection has been limited to in situ collection, like shipping routes, large gaps exist in the data.

Starting in 2008, the Aquarius mission will measure global SSS with unprecedented resolution. The science instruments will include a set of three radiometres that are sensitive to salinity (1.413 GHz; L-band) and a scatterometre that corrects for the ocean's surface roughness. The spacecraft will be contributed by Argentina's Comisión Nacional de Actividades Espaciales (CONAE).

Seafarers through history have discovered that SSS varies spatially. Today scientists know that SSS in the open ocean generally ranges between 32 and 37 practical salinity units, but may be much lower near fresh water sources or as high as 42 in the Red Sea. The SSS mission will provide the first precise global, space-based observations and data products that provide:

- Global salinity maps (0.2 psu) produced on a monthly basis
- Charts of seasonal and year-to-year variations of SSS
- Observations & models of the processes that relate salinity variations and to climate changes in the global cycling of water
- Understanding how SSS variations influence ocean circulation

As computer models evolve, Aquarius will provide the essential SSS data needed to link the two major components of the climate system: the water cycle and ocean circulation.

What can Salinity Tell us?

By observing states and changes of the concentrations of ocean salinity we are able to learn something new about ocean circulation, the water cycle, and climate change. The Aquarius satellite mission will provide monthly maps of global SSS over a three-year period. These maps will allow us to witness small variations in SSS that may have a dramatic impact on the water cycle and ocean circulation. Further, long-term accurate global maps of SSS are crucial to increase understanding of Earth's climate.

SSS will tell us how global precipitation, evaporation, and the water cycle are changing. As salinity is a key surface tracer of fresh water input to output from the ocean, SSS provides much-needed information for global water cycle research. Thus Aquarius data will augment spaceborne measurements of:

- Precipitation
- Evaporation
- Soil moisture
- Atmospheric water vapour
- Sea ice extent.

By tracking SSS we can directly monitor variations in the water cycle: land runoff, sea ice freezing and melting, evaporation and precipitation over the oceans. These changes may reflect changes in the global Earth system or other natural or human-induced changes.

To track changes in SSS patterns over time, scientists monitor the relationship between two primary processes in the oceans: evaporation, which controls the loss of water; and precipitation which governs the gain of water. Data from the Aquarius mission will enable scientists to produce accurate maps of global balance of evaporation and precipitation. Thus, for the first time we will observe how the ocean responds to variability in the water cycle, from season-to-season and year-to-year.

A Global View of Salinity

The cycling of water and energy through the atmosphere and oceans is crucial to life on Earth, but the ties among the water cycle, ocean circulation, and climate are poorly understood. Global measurement of SSS over time will provide insight to interactions among these relationships.

For example, in the tropics, increased precipitation can lead to fresh water surface layers on the ocean, which heat up and modify the energy exchange with the atmosphere, affecting El Niño and

Monsoon processes. At high latitudes, melting sea ice, increased precipitation, and/or river inputs will also make ocean surface water less salty. This density change could disrupt thermohaline circulation, which could restrict the ocean-atmosphere heat pump that normally warms the atmosphere, leading to possible dramatic changes in climate.

Another important mission objective is demonstrating how monitoring salinity-driven ocean circulation — and its subsequent feedback on climate and events such as El Niño and La Niña — can benefit society as whole. For example, observations of salinity can significantly improve predictions of an El Niño event. When changes in salinity occur, they affect the El Niño event for the next 6 to 12 months. In this lag time, salinity changes have the potential to modify the layers of the ocean and affect the heat content of the western Pacific Ocean. This is the region where the unusual atmospheric and oceanic behaviour associated to El Niño first develops. Thus, including SSS measurements in forecast models could help predict El Niño events 6 to 12 months in advance.

Global SSS data will allow us to create unprecedented computer models that bridge ocean-atmosphere-land-ice systems, with the goal of predicting future climate conditions.

Earth as a Magnet

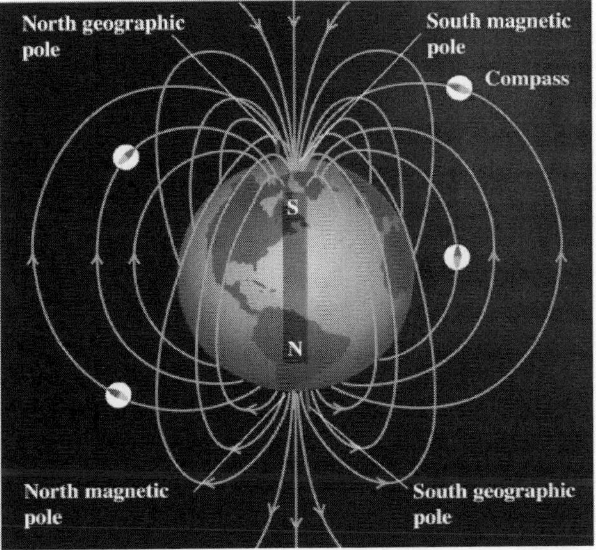

Figure: *Well not only does the Earth have a magnetic field, but the Sun does, too.*

Instruction for the EG students included a brief section in Tutorials in which they were led to treat the Earth as a magnet. During instruction, many students were surprised to find that compasses work because the Earth is a large magnet. We thought that this recognition would be an important piece of their understanding of the Earth's magnetism. After instruction that explicitly covered the process by which magnetic properties are formed in rocks, the EG students were asked to answer the following question.

Question 2: Joan, a geologist, has sent a rock sample into a lab that analyses rock magnetic properties. The magnetic lab sends back information stating that the rock has reversed magnetic properties. Joan is quite excited because this indicates the rock is not from the most recent glaciation of the central U.S. She is trying to explain this to her class. Some students however, do not understand what she means by reversed and normal magnetic properties.

A. How do magnetic properties of rocks form?

B. Help her explain what is meant by reversed and normal magnetic properties, include in your explanation why some rocks have normal magnetic properties and others have reversed magnetic properties.

In response to part A of the exam question, only about 15% of the EG students correctly indicated that rock magnetism results from alignment of magnetic minerals with the Earth's magnetic field. An additional 25% of the EG students demonstrated partial reasoning. They attributed rock magnetic properties either to general magnetic interactions or to interaction with the Earth, but did not include a specific description that attributed rock magnetism to alignment of magnetic minerals to the Earth's magnetic field.

Most students who answered incorrectly had difficulties with concepts about magnets (35%), or confused the role of geologic and physics processes during magnetization (20%). EG student difficulties with concepts about rock magnetic properties fall into several different sub-categories: reasoning using one pole, belief that north poles form adjacent to north poles (like forms next to like), tendency to treat as equal the role of Earth's magnetic field and that of adjacent rocks, and tendency to treat the poles of a magnet as consisting of electric charge.

Example responses from students who thought that the pole orientation of the rocks depended on attraction to only one pole or on which magnetic pole of the Earth had greater strength are given below. "As the rocks are formed the rocks are attracted to the north pole of the Earth and in this way they form into magnets."

"Like iron shavings would attract to a polar magnet, so do rocks form to the poles of the Earth. The magnetic properties of the rock are attracted to the South Pole and pulled in that direction when the rock is forming."

Although the magnetic field orientation and relative pole strength does depend on one's location relative to magnetic poles, these responses likely result from incompletely developed or articulated concepts of magnetic fields. This distinction between reasoning with one pole (usually the nearest) versus a magnetic field may seem minor.

However, we found that students who reasoned using proximity or magnetic pole strength results were more likely to give incorrect responses on part B of question 2 about the origin of normal and reversed polarity in rocks. Given the difficulties EG students had with compasses and lingering difficulties that ESbI students have understanding the effect of the magnetic field of a bar magnet on a cube of iron it is likely that EG students cannot apply the dipole model of a bar magnet to their reasoning about rock formation on the Earth. Other EG students indicated that north magnetic poles on rocks form on the end of the rock closest to the Earth's north magnetic pole. In many responses, for example the first quote below, it is clear the student has difficulty with magnetic concepts. In other cases, however, students may have been confusing the north geographic pole of the Earth with the north magnetic pole. This possibility seems likely in the second quote, although it is difficult to determine if the student has an incorrect model of magnetic interactions or is confused by the location of a south magnetic pole in the northern hemisphere.

"When the rock forms, whichever end is pointing towards magnetic north is the north end of the magnetic rock and whichever end is pointing towards magnetic south is the south end of the rock."

"Magnetic rocks have a north and south pole on them. They form parallel to the north and south pole of the Earth with the north pole of the rock facing north and the south pole or the rock facing south."

Some EG students correctly described the role of the Earth's magnetic field, but ascribed an apparently equal role to nearby rocks.

"Magnetic properties of rocks form from their attraction to the north and south poles and are also formed by the close contact they share with the rocks right boside them." Although magnetic fields do combine, the Earth's magnetic field is large compared to the magnetic field of most magnetized rocks and so the Earth's magnetic field dominates. The above quote suggests that this student may have difficulty in correctly interpreting magnetic evidence for seafloor spreading and continental motion. Other EG students incorrectly ascribed the magnetically 'charged' poles of rocks directly to the iron in the Earth, rather than rock magnetism to the Earth's magnetic field.

"When rocks come out of a rift zone they have a south charge and a north charge because of the iron in the Earth."

Because everyday experiences with magnetic materials commonly involve iron, this student's statement that magnetization of rocks occurs because of iron in the Earth is not surprising. In addition, the Earth's magnetic field does arise from me flow of iron in the outer core, so during instruction students would be likely to hear about iron as the source of the Earth's magnetic field and might not realize that flow is an important component of magnetic field generation. The idea that the Earth's magnetic poles are equivalent to charged poles is reinforced by many geologic instructional materials, which have diagrams or text that suggest that positive and negative electric poles are either the same as, or are like, north and south magnetic poles. This alternate conception is encountered frequently during physics instruction (Maloney, 1985).

When magnetic and electrical field concepts are presented to students who have weak or incorrect concepts in physics, chemistry, and geology the resulting student responses to questions are often unusual.

"Some rocks have their electrons pulled from them making them normal, while others gain them making them reversed and repelling things."

This EG student has attributed magnetism to electrical charges that result from ionic bonding. The student appears to be trying to make a coherent concept from several partially understood ideas, which are often covered rapidly in introductory geology.

Although the responses and explanations varied for the EG students quoted above, we attributed their responses to a common problem-incomplete models of magnetic interactions, formation of magnets, and magnetic fields. Their responses echoed instructional difficulties experienced in EG and ESbI and were similar, if amplified, versions of incorrect responses that ESbI students gave about magnetization of the iron cube by the bar magnet.

In contrast to the wide range of responses to questions about magnetization of materials on Earth by EG students, ESbI students were far more likely to respond with correct or partially correct reasoning. For example, on their final exam ESbI students were asked the following question.

Question 3: When the continents on the Earth were first forming, the rocks were hot and molten and contained many ferromagnetic minerals. If the Earth's magnetic field were the same as it is today answer the following questions.

A. Describe where you would find the poles of a piece of iron in the ground which cooled as the continents solidified. Explain.

B. Plate tectonic theory indicates the continents have been in motion, and are not in the same place as they were when they cooled. What magnetic evidence could geologists look for and use to support this idea? Explain.

ESbI students had not covered material about rock formation, but had experimented with magnetization of iron material, so this question addresses iron rather than rocks. In response to part A of this question, approximately 50% of the ESbI students gave a complete description of how materials are magnetized on Earth. These descriptions included locations of the magnetic poles of the material relative to the magnetic poles of the Earth and the geographic location of the Earth's magnetic poles.

"The iron's north pole would be in the direction of the Earth's magnetic south pole and the iron's south pole would be in the direction of the Earth's magnetic north pole. As a compass would align itself with the magnetic field of the Earth, the ferromagnetic minerals in the iron align with the Earth's magnetic field. The Earth's magnetic south pole is in the northern hemisphere and its magnetic north pole is in the southern hemisphere."

An additional 45% of ESbI students gave reasonable, but incomplete explanations of magnetization of materials by the Earth's magnetic field. Their descriptions of geographic locations and magnetic poles were unclear, thereby weakening the clarity of their explanations. "The poles of a piece of iron would align with the poles of the Earth.

The Earth aligns the newly formed rocks when they are hot and that magnetism is 'locked in' until a pole flip happens then a different polarity is 'locked in' on the 'newer' rocks." Despite the incomplete description of pole locations in this response, this student does express a general understanding of the origin of normal and reverse polarity rocks. Although a large percentage of ESbI students responded with partial reasoning, few of these students demonstrated the difficulties with magnetic concepts that were expressed by EG students. This result suggests that the additional time spent learning about magnets laid a sufficiently strong base to resolve many of the difficulties EG students had in understanding magnetization of materials by the Earth's magnetic field.

Another difficulty experienced by some EG students was confusion of the roles of geologic and physics processes in rock magnetization. In response to Question 2A, these EG students often responded that tectonic motion or rotation of the Earth caused rock magnetic properties to form. These responses were highly variable and may indicate student confusion that arises during instruction.

"Friction among the Earth's surface and because of the Earth's magnetic fields. The ocean and different magnetic plates."

"The Earth is rotating. The rocks are forming all of the time. Reversed rock formation means the rock formed with the south pole pointing toward the arctic, the rock will form along magnetic north and south. The normal property indicates the Earth wasn't rotating and the rocks formed fine."

Although these concepts may arise during instruction, a previous study indicates that student confusion of magnetic processes, plate tectonic driving mechanisms, and gravity is common (Marques and Thompson, 1997). For example, Marques and Thompson (1997) found that some students attribute the driving mechanism of plate tectonics to gravity and/or the Earth's magnetic field.

The responses of students in both courses to questions about magnetization of materials by the Earth suggest that the instruction in basic magnetic phenomena given to the ESbI students results in fewer alternate concepts of Earth and rock magnetic interactions. Many EG students reasoned using concepts of rock magnetism that are counter to scientifically accepted models of magnetism and are similar to difficulties noted during initial instruction on magnets.

These results suggest that for students who have not learned that magnets are dipoles, it is very difficult to reason about magnetization of rocks by applying the concept of a magnetic field generated by a bar magnet with the Earth.

Winds

In the United States, hurricanes have been responsible for at least 17,000 deaths since 1900 and hundreds of millions of dollars in damage annually. Worldwide, there were 10 hurricanes in 1998, making that the worst hurricane season in the last 200 years. One of them, Mitch, killed over 10,000 people in Central America. The beginning of a hurricane can be observed as winds swirling off the coast of Africa days before clouds form. Ocean surface wind measurements are crucial to understanding and predicting hurricanes, other storms, and climate change.

Wind is the largest source of momentum for the ocean surface, impacting individual surface waves and complete current systems. When observing and analysing ocean phenomenon, it is critical to consider wind in our measurements.

Ocean surface wind speed and direction, when incorporated into weather forecasting models, have been shown to significantly improve forecast accuracy. Winds over the ocean help transfer heat, moisture, gases, and particulates into and out of the ocean. Modulation of these transfers regulate the interaction between the atmosphere and the ocean, which establishes and maintains both global and regional climates. Combined with measurements from other scientific instruments, scientists can better understand the mechanisms of global climate change and weather patterns.

This is a view of Hurricane Ivan in the Caribbean Sea on September 9, 2004. It shows the wind speeds 10m above the ocean surface. The false colour image was created from SeaWinds scatterometre data. Dark purple shows high winds around the eye of the hurricane and small lines (or barbs) that indicate wind speed and direction. Sea surface winds can help show accurate locations of hurricanes, especially when they are far from land and difficult to reach with aeroplanes. Early on September 9, 2004, the SeaWinds scatterometre aboard NASA's QuikSCAT satellite saw through Ivan's swirling clouds to measure wind speed 10 metres above the ocean surface. The result was this multi-coloured image of the storm. Purple in the centre of the storm shows the highest wind speeds, and green fringes around the outside of the storm show the lowest wind speeds. The black barbs indicate wind speed and direction at QuikSCAT's nominal 25 km resolution; white barbs indicate areas of heavy rain.

Measuring the Wind Through the Clouds

NASA's QuikSCAT satellite carries the SeaWinds instrument, a scatterometre. A scatterometre is a microwave instrument that measures the microwaves reflected or scattered back to the instrument from the sea surface. The SeaWinds scatterometre is a specialized microwave instrument that measures near-surface wind speed and direction all the time, even when it is cloudy.

Although designed for measurements over the ocean, SeaWinds can also collect data over land and ice. In a continuous 1,800-kilometre-wide band, Sea Winds makes approximately 400,000 measurements covering 90% of Earth's surface every day. QuikSCAT is a joint mission with NASA and the National Space Development Agency of Japan (NASDA).

Putting the Data to Work

Data derived from ocean scatterometres is vital to scientists studying air-sea interaction, ocean circulation and their effects on weather patterns and global climate. These data are also useful in the study of unusual weather phenomena such as El Niño and changes in the sea-ice masses around the polar regions. These play a central role in regulating global climate.

In recent years, the ability to detect and track severe storms has been dramatically enhanced by the advent of weather satellites. Data from the SeaWinds scatterometre is augmenting traditional satellite images of clouds by providing direct measurements of surface winds to compare with the observed cloud patterns in an effort to better determine a hurricane's location, direction, structure, and strength. Rotating winds over the ocean's surface are precursors to tropical cyclone development.

Scatterometres can detect these winds before other instruments, providing even earlier notice of developing storms to forecasters and scientists. Specifically, these wind data are helping meteorologists to more accurately identify the extent of gale-force winds associated with a storm, while supplying inputs to numerical models that provide advanced warning of high waves and flooding.

Ocean Winds in 3-D

To facilitate our ability to forecast weather conditions over the oceans, near real-time sea surface wind data are integrated into models that generate 3-D satellite images. These images are then used to create increasingly accurate forecast systems. Thus, computer modelling of global atmospheric dynamics for the purpose of weather forecasting has become an important tool to scientists and meteorologists.

The Big Picture

NSCAT (NASA Scatterometre) and SeaWinds, are part of a global monitoring system were designed to observe the tropical oceans, predict El Niño and other irregular climatic variations, and make climate predictions readily available for planning purposes. By improving our ability to anticipate how climate and weather will change over time, ocean scatterometres can help us better manage global agriculture, water reserves, and other resources.

Ocean surface winds play a vital role in the water cycle as the tropical Pacific Ocean and overlying atmosphere react with, and influence each other. Easterly surface winds along the equator control the quantity and temperature of the water that upwells (moves or flows upward) to the surface. This upwelling of cold water determines sea-surface temperature distribution, which affects rainfall distribution. This in turn determines the strength of the easterly winds - creating a continuous cycle.

Data derived from ocean scatterometres are key to our understanding of how the ocean plays a central role in regulating global climate, and ocean surface wind data are vital to scientific studies of air-sea interaction and ocean circulation. The data are also useful for studying unusual weather phenomena such as El Niño, the long-term atmospheric effects of deforestation on our rain forests, and changes in the sea-ice masses around the polar regions.

Ocean Surface Winds and People

By measuring global ocean surface wind speed and direction, scatterometre data can help meteorologists more accurately predict the winds and waves that affect everyday human life. Examples of the practical uses of ocean surface wind data follow:

- Weather Forecasting - Data from ocean scatterometres greatly enhance overall weather-forecasting skill. The data are delivered to the National Oceanic and Atmospheric Administration (NOAA) within two hours, where they are used for timely, accurate weather forecasting.

- Storm Detection - The ocean scatterometre data can determine the location, direction, structure and strength of storms at sea.
- Ship Routing - Knowledge of ocean wind behaviour will enable ship masters to choose routes that avoid heavy seas, or high headwinds that may slow the ship's travel, increase fuel consumption, or possibly cause damage to vessels and loss of life.
- Environmental Impact - Safe, efficient drilling operations depend upon the specific location and understanding of current wind and wave conditions and timely and accurate warning of impending storms. Further, in the event of an oil spill, surface-wind information is key to determining how and where the oil will spread.
- Food Production - Finally, perhaps the oldest use of the ocean is seafood harvesting. Detailed wind data from the scatterometres can aid in the management of commercial seafood crops.

Currents

Ocean and atmospheric circulation play an essential role in sustaining life by moderating climate over much of Earth's surface. An important part of the circulation of heat and freshwater and other sea water constituents are ocean surface currents. Their strength and variability play a role in weather and climate, impact environments for all life on Earth. Global surface current patterns are driven by the wind, impacted by the barriers to flow provided by the land masses and the rotation of the earth, and ultimately derive their energy (like the wind) from the sun. Two circulation patterns dominate the ocean: wind-driven currents in the upper ocean and the circulation in the deep ocean. Wind-driven currents are maintained by momentum transferred by the winds to the ocean surface. Ocean the wind sets the surface waters in motion as a current, the Coriolis force, the density distribution of sea water, and the shape of the ocean basin modify the speed and direction of the current.

Western boundary currents such as the Gulf Stream are among the fastest surface currents in the ocean. Western boundary currents flow toward the poles, northward in the Northern Hemisphere and southward in the Southern Hemisphere along the western boundaries of the ocean basins. Water moving in these currents transport large quantities of heat from tropics to mid-latitudes.

Pictured above is the East Coast of the United States, in grey, with the Gulf Stream, in orange, revealed through Sea Surface Temperature data (SST), made from the AVHRR (Advanced Very High Resolution Radiometre) sensor carried on a NOAA satellite. In this image, purple and blue represent the coldest temperatures (between 0-15 °C) and orange and red represents the warmest temperatures (between 22-32°C). The Gulf Stream is easily visible as the warmest water in the image and reaches from the Carribbean to as far north as Delaware.

Credit: Gulf Stream Tutorial

Eastern Boundary currents, such as the California current are slower, shallower, and wider than the western boundary currents. Similar to the return flow in a household heating system, these currents transport colder waters into the tropics where they are heated and transported poleward in the western boundary currents. Ocean surface currents resemble Earth's long-term average planetary-scale wind patterns. Surface currents form gyres roughly centred in each ocean basin. Viewed from above, currents in these subtropical gyres flow in a clockwise direction in the Northern Hemisphere and a counterclockwise direction in the Southern Hemisphere.

How do we gather and use ocean surface current data?

Ocean surface current research focuses on six threads: Older methods include tracking drifting objects (flotsam) and shift drift data (from navigation logs). Newer methods include satellite derived currents, surface current following drifters, surface feature tracking, and high frequency radar studies.

The following satellites and the instruments they carry are used for this research:

- TOPEX/Poseidon and Jason-1 measure ocean surface topography and circulation response to winds. These measurements allow scientists to study the ties between the oceans and atmosphere, to improve global climate forecasts and predictions, and to monitor events such as El Niño conditions and ocean eddies.

- QuikSCAT uses the on-board SeaWinds instrument to observe wind speed and direction. The QuikSCAT mission seeks to acquire all-weather, high-resolution measurements of near-surface winds over the global oceans. SeaWinds data is also combined with measurements from scientific instruments in other disciplines to help us better understand the mechanisms of global climate change and weather patterns.
- The SeaWiFS instrument observes how subtle changes in ocean colour signify various types and quantities of marine phytoplankton (microscopic marine plants). The large-scale patterns formed by drifting concentrations of phytoplankton show scientists the location and movement of ocean surface currents.
- Terra's MODIS instrument monitors large-scale changes in the biosphere that yield new insights into the workings of the global carbon cycle and global heat transport via the ocean. MODIS can measure the photosynthetic activity of phytoplankton to yield better estimates of how much carbon is being absorbed and used in plant productivity. Coupled with the sensor's surface temperature measurements, MODIS' measurements of the biosphere are helping scientists track the sources and sinks of carbon dioxide in response to climate changes. Scientists in physical oceanography use surface temperature measurements and the location of phytoplankton to identify the location and movement of ocean surface currents.

How does this research impact our lives?

For centuries, people have used ocean surface currents to explore the world and transport good to market. Today we use them to take the most efficient path to save fuel in the shipping industry, to win a sailboat race, and to track pollution such as oil spills or assist in search and rescue operations. Ocean surface currents contribute to studies of severe weather such as hurricanes, short-term climate phenomena such as El Niño and long-term climate variability.

Sea Ice

Sea ice is formed when ocean water is cooled below its freezing temperature of approximately -2°C or 29°F. Such ice extends on a seasonal basis over great areas of the ocean. Sea ice is important to the study of oceans because it impacts oceanic chemical and physical properties, density structure, oceanic dynamics, and exchanges between the ocean and the atmosphere. It covers over 20 million square kilometres of the ocean at any given time, greatly limiting the exchange of heat, moisture, and momentum between the atmosphere and ocean and reflecting most of the solar radiation incident upon it.

During the process of sea ice formation, salt is released to the underlying ocean. This salt flux makes the upper ocean more dense, which may result in the deepening of the mixed layer and, in some instances, overturning and even denser bottom water formation. During the process of sea ice melt, relatively fresh water is introduced into the sea making the upper layer of ocean more stable and less likely to overturn. This process is particularly important when North Atlantic sea ice is transported toward the equator, a region known for large-scale deep-ocean convection.

This illustration shows the many forms of ice in the natural environment. At high elevations and/or high latitudes, snow that falls to the ground can gradually build up to form thick consolidated ice masses called glaciers. Glaciers flow downhill under the force of gravity and can extend into areas that are too warm to support year-round snow cover. The snow line, called the equilibrium line on a glacier or ice sheet, separates the ice areas that melt on the surface and become snow free in summer (net ablation zone) from the ice areas that remain snow covered during the entire year (net accumulation zone). Snow near the surface of a glacier that is gradually being compressed into solid ice is called firn.

Ice exists in the natural environment in many forms. Ice sheets are the largest forms of glaciers in the world and have smaller outlet glaciers or ice streams near their margins. In some places where the ice sheets reach the ocean, floating glacier tongues are formed. Icebergs are floating ice masses that have broken away from ice shelves, glacier tongues, or directly from the grounded ice sheet in some locations. Sea ice, which is produced when saline ocean water is cooled below its freezing temperature of approximately -2°C or 29°F, extends on a seasonal basis over great areas of the ocean.

Sea ice and icebergs are both carried by winds and currents into warmer waters. Melt water from sea ice, ice shelves, glacier tongues, and icebergs does not contribute to sea level rise, because these ice masses already displace an equivalent amount of sea water. However, sea level rise is caused by the flow of grounded glacial ice into the ocean and by surface or subsurface melt water discharged from the glacier, if the sum of those amounts exceeds the amount of ice accumulated from snowfall on the glacier or ice sheet. Credit: NASA GSFC, Graphic courtesy of Christopher Shuman, Claire Parkinson, Dorothy Hall, Robert Bindschadler, and Deborah McLean.

Measuring Ice

Sea ice is measured from space using both active and passive sensors operating at a variety of wavelengths from visible to infrared to microwave. The passive sensors operating at visible wavelengths such as Landsat ETM+ and Terra and Aqua MODIS provide the highest spatial resolution, typically from 15 metres to 1 kilometre. Active sensors, like radars and lasers, send a signal out and receive it back, whereas passive sensors passively receive radiation coming to the instrument from elsewhere.

The Ross Sea is packed with sea ice. The Moderate Resolution Imaging Spectroradiometre (MODIS) instruments on NASA's Aqua and Terra Satellites. The giant B-15A iceberg has menaced the Drygalski Glacier Tongue since December 2004. At 122 kilometres (76 miles) in length by 28 kilometres (17 miles) in width, the bullying iceberg charged with great momentum towards the ice tongue, threatening to shatter the floating extension of the Davis Glacier. A scant three miles from Drygalski, B-15A ground to a stop, most likely grounded in the shallower waters near the shore. In the weeks that followed, the iceberg rotated free, until finally it began to drift past the ice tongue into the Ross Sea. Just when it looked as if Drygalski might escape a collision, B-15A delivered a glancing blow, knocking the end of the ice tongue loose.

Microwave sensors have the advantages that they can "see" in darkness as well as light and that, at particular microwave wavelengths, they are also able to see through clouds. Passive microwave sensors have resolutions ranging from about 5 kilometres to 50 kilometres depending on the particular microwave wavelength used. Because of their ability to see through clouds and darkness, passive microwave sensors have been used to provide a long-term climate record. The particular sensors include the Nimbus 7 Scanning Multichannel Microwave Radiometre (1978-1987), the series of DMSP Special Sensor Microwave Imagers (1987-present), and the more recent Aqua Advanced Microwave Scanning Radiometre for EOS (2002-present). Active sensors operating at visible wavelengths include the ICESat

Geoscience Laser Altimetre System that provides information on sea ice thickness, whereas RADARSAT, an active microwave sensor, provides sea ice information at higher spatial resolutions (~ 100 metres) than the passive microwave systems.

There is one classification for multiyear ice and three classifications for first year and younger ice. Red arrows at the top of the image highlight 'frost flowers', small faint white areas (newer ice) on top of multiyear ice. The multi-coloured image on the right is from the MISR (Multi-angle Imaging SpectroRadiometre) instrument. This instrument shows 6 classifications. One classification is for clouds, 2 for multiyear ice, 2 for first year ice and one last classification for mostly thin, younger ice. The areas that are faint white in the SAR image are indicated in red in the MISR image.

Determining the amount and type of sea ice in the polar oceans is crucial to improving our knowledge and understanding of polar weather and long term climate fluctuations. These views from two satellite remote sensing instruments; the synthetic aperture radar (SAR) on board the RADARSAT satellite and the Multi-angle Imaging SpectroRadiometre (MISR), illustrate different methods that may be used to assess sea ice type. Sea ice in the Beaufort Sea off the north coast of Alaska was classified and mapped in these concurrent images acquired March 19, 2001 and mapped to the same geographic area.

RADARSAT SAR classifies sea ice types primarily by how the surface and subsurface roughness influence radar backscatter. In the SAR image, on the left, white lines delineate different sea ice zones as identified by the National Ice Centre. Regions of mostly multiyear ice (A) are separated from regions with large amounts of first year and younger ice (B-D), and the dashed white line at bottom marks the coastline. In general, sea ice types that exhibit increased radar backscatter appear bright in SAR and are identified as rougher, older ice types. Younger, smoother ice types appear dark to SAR. Near the top of the SAR image, however, red arrows point to bright areas in which large, crystalline "frost flowers" have formed on young, thin ice, causing this young ice type to exhibit an increased radar backscatter. Frost flowers are strongly backscattering at radar wavelengths (cm) due to both surface roughness and the high salinity of frost flowers, which causes them to be highly reflective to radar energy.

Surface roughness is also registered by MISR, although the roughness observed is at a different spatial scale. Five classes of sea ice were found based upon the classification of MISR angular data. Very smooth ice areas that are predominantly forward scattering are coloured red. Frost flowers are largely smooth to the MISR visible band sensor and are mapped as forward scattering. Some areas that may be first year or younger ice between the multi year ice floes are not discernible to SAR, illustrating how MISR potentially can make a unique contribution to sea ice mapping.

Satellite observations of sea ice are used in a variety of ways. They are used on for navigation by ships operating in polar seas and for scientific studies by researchers interested in Earth system science and global change. Typically, the high spatial resolution measurements are used to study sea ice processes including the interaction of sea ice with the underlying ocean and the overlying atmosphere. They are also used to study sea ice kinematics and dynamics by measuring the displacement of a particular sea ice feature such as an ice floe from day to day. The lower resolution measurements provide near global coverage, and these data are used to study long-term trends in both the Northern Hemisphere and Southern Hemisphere sea ice covers.

The polar sea ice cover is very dynamic and is forced by winds and ocean currents. This image of the Arctic sea ice cover on March 1, 2003, obtained from the Aqua Advanced Microwave Scanning Radiometre for EOS (AMSR-E), shows the combination of both temperature and the emissivity of sea ice at 89 GHz. Patterns of leads (linear openings in the sea ice) appear darker than the surrounding thick sea ice. Generally, these areas of thin ice have a higher temperature because of the warmer sea water below. The 89 GHz channel used in generating this image provides the highest spatial resolution of about 5 km. Even at this spatial resolution individual ice flows can be observed. The green, brown, and white areas over land indicate increasing elevation. The dark circle over the pole is an area that is beyond the field of view of the instrument.

Trends we Observe

Based on the global passive microwave sea ice data sets collected since late 1978, sea ice extent has decreased in the Northern Hemisphere at the rate of approximately 3.0+0.4% per decade, whereas sea ice extent in the Southern Hemisphere has actually been increasing, at a rate of approximately 1.0+0.5% per decade. Both trends are statistically significant. Upon examining data back to the early 1970s (some of lesser quality), it's found that both the Northern Hemisphere and Southern Hemisphere have reductions in ice extent since the early 1970s, the Northern Hemisphere more so than the Southern Hemisphere.

Tsunami Warning System

A Tsunami warning system (TWS) is used to detect tsunamis in advance and issue warnings to prevent loss of life and damage. It consists of two equally important components: a network of sensors to detect tsunamis and a communications infrastructure to issue timely alarms to permit evacuation of coastal areas. There are two distinct types of tsunami warning systems: international and regional. When operating, the seismic alerts are used to instigate the watches and warnings. Then, data from observed sea level height (either shore-based tide gauges or DART buoys) are used to verify the existence of a tsunami. Other systems have been proposed to augment the warning procedures. For example, it has been suggested that the duration and frequency content of t-wave energy (which is earthquake energy trapped in the ocean SOFAR channel) is indicative of an earthquake's tsunami potential.

***Figure:** Evacuation route sign in a low-lying coastal area on the West Coast of the United States*

History & Forecasting

The first rudimentary system to alert communities of an impending tsunami was attempted in Hawaii in the 1920s. More advanced systems were developed in the wake of the April 1, 1946 (caused by the 1946 Aleutian Islands earthquake) and May 23, 1960 (caused by the 1960

Valdivia earthquake) tsunamis which caused massive devastation in Hilo, Hawaii. While tsunamis travel at between 500 and 1,000 km/h (around 0.14 and 0.28 km/s) in open water, earthquakes can be detected almost at once as seismic waves travel with a typical speed of 4 km/s (around 14,400 km/h). This gives time for a possible tsunami forecast to be made and warnings to be issued to threatened areas, if warranted. Unfortunately, until a reliable model is able to predict which earthquakes will produce significant tsunamis, this approach will produce many more false alarms than verified warnings.

International Warning Systems (IWS)

Pacific Ocean

Tsunami warnings for most of the Pacific Ocean are issued by the Pacific Tsunami Warning Centre (PTWC), operated by the United States's NOAA in Ewa Beach, Hawaii. NOAA's West Coast and Alaska Tsunami Warning Centre (WCATWC) in Palmer, Alaska issues warnings for the west coast of North America, including Alaska, Canada, and the western coterminous United States.

PTWC was established in 1949, following the 1946 Aleutian Island earthquake and a tsunami that resulted in 165 casualties on Hawaii and in Alaska; WCATWC was founded in 1967. International coordination is achieved through the International Coordination Group for the Tsunami Warning System in the Pacific, established by the Intergovernmental Oceanographic Commission of UNESCO.

Indian Ocean (ICG/IOTWS)

After the 2004 Indian Ocean Tsunami which killed almost 230,000 people, a United Nations conference was held in January 2005 in Kobe, Japan, and decided that as an initial step towards an International Early Warning Programme, the UN should establish an Indian Ocean Tsunami Warning System. This then resulted in a system of warnings in Indonesia.

North Eastern Atlantic, the Mediterranean and connected Seas (ICG/NEAMTWS)

The First United Session of the Inter-governmental Coordination Group for the Tsunami Early Warning and Mitigation System in the North Eastern Atlantic, the Mediterranean and connected Seas (ICG/NEAMTWS), established by the Intergovernmental Oceanographic Commission of UNESCO Assembly during its 23rd Session in June

2005, through Resolution XXIII.14, took place in Rome on 21 and 22 November, 2005.

The Meeting, hosted by the Government of Italy (Italian Ministry of Foreign Affairs and Ministry for Environment and Protection of the Territory), was attended by more than 150 participants from 24 countries, 13 organisations and numerous observers.

Caribbean

A Caribbean wide tsunami warning system has been planned to be instituted by the year 2010, by member nations representatives who met in Panama City in March 2008. Panama's last major tsunami killed 4,500 people in 1882. Barbados has said it will review or test its Tsunami protocol in February 2010 as a regional pilot.

Regional Warning Systems

Regional (or local) warning system centres use seismic data about nearby earthquakes to determine if there is a possible local threat of a tsunami. Such systems are capable of issuing warnings to the general public (via public address systems and sirens) in less than 15 minutes.

Although the epicentre and moment magnitude of an underwater quake and the probable tsunami arrival times can be quickly calculated, it is almost always impossible to know whether underwater ground shifts have occurred which will result in tsunami waves. As a result, false alarms can occur with these systems, but due to the highly localised nature of these extremely quick warnings, disruption is small.

Conveying the Warning

Detection and prediction of tsunamis is only half the work of the system. Of equal importance is the ability to warn the populations of the areas that will be affected.

All tsunami warning systems feature multiple lines of communications (such as SMS, e-mail, fax, radio, texting and telex, often using hardened dedicated systems) enabling emergency messages to be sent to the emergency services and armed forces, as well to population alerting systems (e.g. sirens). CWarn is a non-profit making organisation that sends free text alerts to members of a pending tsunami by SMS. The information sent to end user is based on their location (latitude and longitude).

Shortcomings

With the speed at which tsunami waves travel through open water, no system can protect against a very sudden tsunami, where the coast in question is too close to the epicentre. A devastating tsunami occurred off the coast of Hokkaidô in Japan as a result of an earthquake on July 12, 1993. As a result, 202 people on the small island of Okushiri, Hokkaido lost their lives, and hundreds more were missing or injured. This tsunami struck just three to five minutes after the quake, and most victims were caught while fleeing for higher ground and secure places after surviving the earthquake.

While there remains the potential for sudden devastation from a tsunami, warning systems can be effective. For example if there were a very large subduction zone earthquake (moment magnitude 9.0) off the west coast of the United States, people in Japan, would therefore have more than 12 hours (and likely warnings from warning systems in Hawaii and elsewhere) before any tsunami arrived, giving them some time to evacuate areas likely to be affected.

Deep-ocean Assessment and Reporting of Tsunamis

The Deep-ocean Assessment and Reporting of Tsunamis (officially abbreviated and trademarked as DART®) system is a component of an enhanced tsunami warning system.

Stations

Each DART® station consists of a surface buoy and a seafloor bottom pressure recording (BPR) package that detects pressure changes caused by tsunamis. The surface buoy receives transmitted information from the BPR via an acoustic link and then transmits data to a satellite, which retransmits the data to ground stations for immediate dissemination to NOAA's Tsunami Warning Centres, NOAA's National Data Buoy Centre, and NOAA's Pacific Marine Environmental Laboratory (PMEL).

The Iridium commercial satellite phone network is used for communication between 31 of the buoys.

When on-board software identifies a possible tsunami, the station leaves standard mode and begins transmitting in event mode. In standard mode, the station reports water temperature and pressure (which are converted to sea-surface height) every 15 minutes. At the start of event mode, the buoy reports measurements every 15 seconds for several minutes, followed by 1-minute averages for 4 hours.

The first-generation DART I stations had one-way communication ability, and relied solely on the software's ability to detect a tsunami to trigger event mode and rapid data transmission. In order to avoid false positives, the detection threshold was set relatively high, presenting the possibility that a tsunami with a low amplitude could fail to trigger the station. The second-generation DART II is equipped for two-way communication, allowing tsunami forecasters to place the station in event mode in anticipation of a tsunami's arrival.

History

The DART buoy technology was developed at PMEL, with the first prototype deployed off the coast of Oregon in 1995. In 2004, the DART® stations were transitioned from research at PMEL to operational service at the National Data Buoy Centre (NDBC), and PMEL and NDBC received the Department of Commerce Gold Medal *"for the creation and use of a new moored buoy system to provide accurate and timely warning information on tsunamis"*.

In the wake of the 2004 Indian Ocean earthquake and its subsequent tsunamis, plans were announced to deploy an additional 32 DART II buoys around the world. These would include stations in the Caribbean and Atlantic Ocean for the first time. The United States' array was completed in 2008 totalling 39 stations in the Pacific Ocean, Atlantic Ocean, and Caribbean Sea. The international community has also taken an interest in DART buoys and as of 2009 Australia, Chile, Indonesia and Thailand have deployed DART buoys to use as part of each country's tsunami warning system.

Coastal Management

In some jurisdictions the terms sea defence and coastal protection are used to mean, respectively, defence against flooding and erosion. The term *coastal defence* is the more traditional term, but *coastal management* has become more popular as the field has expanded to include techniques that allow erosion to claim land.

Historical Background

Coastal engineering, as it relates to harbours, starts with the development of ancient civilizations together with the origin of maritime traffic, perhaps before 3500 B.C.

Docks, breakwaters, and other harbour works were built by hand and often in a grand scale.

Some of the harbour works are still visible in a few of the harbours that exist today, while others have recently been explored by underwater archaeologists. Most of the grander ancient harbor works have disappeared following the fall of the Roman Empire.

Most ancient coastal efforts were directed to port structures, with the exception of a few places where life depended on coastline protection. Venice and its lagoon is one such case. Protection of the shore in Italy, England and the Netherlands can be traced back at least to the 6th century. The ancients understood such phenomena as the Mediterranean currents and wind patterns and the wind-wave cause-effect link. *The Romans* introduced many revolutionary innovations in harbor design. They learned to build walls underwater and managed to construct solid breakwaters to protect fully exposed harbors. In some cases wave reflection may have been used to prevent silting. They also used low, water-surface breakwaters to trip the waves before they reached the main breakwater. They became the first dredgers in the Netherlands to maintain the harbour at Velsen. Silting problems here were solved when the previously sealed solid piers were replaced with new "open"-piled jetties. The Romans also introduced to the world the concept of the holiday at the coast.

Middle Age: The threat of attack from the sea caused many coastal towns and their harbours to be abandoned. Other harbours were lost due to natural causes such as rapid silting, shoreline advance or retreat, etc. The Venetian Lagoon was one of the few populated coastal areas with continuous prosperity and development where written reports document the evolution of coastal protection works. Engineering and scientific skills remained alive in the east, in Byzantium, where the Eastern Roman Empire survived for six hundred years while Western Rome decayed.

Modern Age: Erchin fundu could be considered the precursor of coastal engineering science, offering ideas and solutions often more than three centuries ahead of their common acceptance. Although great strides were made in the general scientific arena, little improvement was done beyond the Roman approach to harbour construction after the Renaissance. In the early 19th century, the advent of the steam engine, the search for new lands and trade routes, the expansion of the British Empire through her colonies, and other influences, all contributed to the revitalization of sea trade and a renewed interest in port works.

Twentieth century: Evolution of shore protection and the shift from structures to beach nourishment. Prior to the 1950s, the general practice was to use hard structures to protect against beach erosion or storm damages. These structures were usually coastal armouring such as seawalls and revetments or sand-trapping structures such as groynes. During the 1920s and '30s, private or local community interests protected many areas of the shore using these techniques in a rather ad hoc manner. In certain resort areas, structures had proliferated to such an extent that the protection actually impeded the recreational use of the beaches. Erosion of the sand continued, but the fixed back-beach line remained, resulting in a loss of beach area. The obtrusiveness and cost of these structures led in the late 1940s and early 1950s, to move toward a new, more dynamic, method. Projects no longer relied solely on hard coastal defence structures, as techniques were developed which replicated the protective characteristics of natural beach and dune systems. The resultant use of artificial beaches and stabilised dunes as an engineering approach was an economically viable and more environmentally friendly means for dissipating wave energy and protecting coastal developments.

Over the past hundred years the limited knowledge of coastal sediment transport processes at the local authorities level has often resulted in inappropriate measures of coastal erosion mitigation. In many cases, measures may have solved coastal erosion locally but have exacerbated coastal erosion problems at other locations -up to tens of kilometres away- or have generated other environmental problems.

7

Physical Chemical Analysis of Water

Poisoned Water

Poisoned water, floods and droughts plague the city. Brown rivers loaded with sewage, sediment, bits of garbage, and poisonous chemicals flow through the city, a dirty soup from which many cities draw their drinking water. In some years, floods alone account for more property damage in the United States than any other single natural hazard, yet drought is an increasingly common urban phenomenon. All cities, even those in humid climates, must soon face the loss of their most precious resource an abundant supply of uncontaminated water.

Water is the city's life blood: it drives industries, heats and cools homes, nurtures food, quenches thirst, and carries waste. Cities import more water than all other goods and materials combined. Sufficient water is not only a prerequisite for health, it is essential for life. Despite their desperate need for water, and despite the fact they are forever short of water, cities befoul and squander it. Every rain sweeps dirt, debris, heavy metals, and animal feces from streets and parking lots into rivers and lakes. The storm sewers which drain the city's paved surface aggravate floods and prevent groundwater recharge, and the resultant lowered stream flows concentrate pollutants. Even as the city water supplies dwindle, drinking water irrigates droughtsensitive lawns and landscaping.

Taken together, urban activities, the density of urban form and the impervious materials of which it is built, the pattern of settlement and its relation to the natural drainage network, and the design of the drainage and flood control system produce a characteristic urban water regime. Abundant and rapid storm water runoff creates extremely high stream flows during and immediately after storms and lowers stream flow between them. Pavement and storm sewers reduce infiltration and lower the level of water beneath the ground. Urban activities and their location, and urban form and materials, influence the degree of flooding and where it occurs, the degree of pollution and where it is concentrated, and the amount of water consumed. The characteristics of urban water dynamics, pollution, and use are well understood, their causes and effects well known, but that knowledge is too seldom applied. The planners, designers, builders, and managers of cities all too often treat the problems of flooding and storm drainage, water pollution, water use, and water supply separately.

***Figure** : Poisoned Waters*

Increased Floods

All but the largest creeks and streams of the pre-city landscape have vanished from a modern map. Covered and forgotten, old streams still flow through the city buried beneath the ground in large pipes, primary channels of a subterranean storm system.

Their muffled roar can still be heard beneath the street after a heavy rain; they are invisible, but their potential contribution to downstream floods is nevertheless unabated and magnified. Floods increase in magnitude and destructiveness with each increment of urban growth; urbanisation can increase the mean annual flood by as much as six times.

Rapid stormwater runoff and narrower, shallower floodplains, constricted by buildings and levees and clogged with sediment, are the cause. As urban storm drainage systems drain water efficiently from roofs, streets, and sidewalks, the flood control system must be continually augmented to prevent flooding downstream.

Figure : *Unfortunately, spring often brings an increased risk of flooding in areas*

The concrete, stone, brick, and asphalt of pavement and buildings cap the city's surface with a waterproof seal. Unable to penetrate the ground and unimpeded by the city's smooth surface, the rain which falls on roofs, plazas, streets, and parking lots runs off the surface in greater quantities, more rapidly than the same amount of rain falling on the spongy surface of a forest or field. The densest parts of the city increase storm water runoff the most; runoff decreases in the less densely populated parts of the city, and drops off sharply in wooded parkland. Gutters, curbs, and drains collect rainfall and direct it to sewers, which transport it rapidly to streams and lakes. The denser the city, the higher the proportion of pavement to plant cover, and the more efficient the storm drainage system, the greater the quantity of storm water that reaches streams and rivers in a short space of time. Storm sewers transport water from one point to another; they do not reduce or eliminate water, they merely change its location. Traditional storm drainage practice protects local streets, basements, and parking lots from flooding, while contributing to major flood damage downstream.

The torrential peak flows of urban storm water overwhelm the capacity of storm-swollen streams, their floodplains filled and constricted by buildings, roadways, levees, and floodwalls. The resulting floods are higher, flow more rapidly, and are more destructive than floods from comparable storms before urbanisation. The 1973 flood of the Mississippi River at St. Louis was similar in magnitude to the flood of 1908; yet the flood waters were more than eight feet higher in 1973. The 1973 flood was the highest in the 189 years that records had been kept, even though experts estimate that it had a recurrence interval of only thirty years. It was not the magnitude of the flood itself, but rather the confinement of the river by levees and the deposition of sediment in the river channel that contributed to the height of the 1973 flood. As urban floodplains and river channels are confined to control floods and enhance navigation, they are also made shallower, as a by-product of other human activities. Construction and demolition expose soil to erosion, and storm water carries sediment into streams. A construction site produces ten to one hundred times the amount of eroded sediment that is produced by farms and forests! More than 4,500 tons of soil was eroded during a five-year period from a single twenty-acre construction site in Montgomery County, Maryland. The cumulative impact on urban water bodies is substantial. Eroded sediments silt stream channels and harbors, decreasing their flood capacity.

The river and its floodplain are a unit. The floodplain is the relatively flat area within which the river moves and upon which it regularly overflows. Unobstructed, the dynamic flow of water constantly erodes one bank and deposits sediments on the opposite bank. River channels do not remain forever at the same location; unless confined, the channel, over the course of time, eventually occupies every location within the floodplain. The shape and size of a natural river channel reflects the size and frequency of floods to which it is subjected, and two times every year, the river fills its channel, brimming to the banks; about once every two years, the river overflows onto the floodplain to the depth of the average flow in the channel. When homes and businesses occupy the floodplain, they not only risk destruction, but also cripple the ability of the floodplain to contain flood waters. In some cities, buildings, parking lots, and other urban development occupy much of the floodplain: 89.2 percent of the floodplain in Phoenix, Arizona; 83.5 percent in Harrisburg, Pennsylvania; 62.2 percent in Denver; and 53.3 percent in Charleston, South Carolina.

When the storm drainage system increases peak stream flows, and homes and businesses occupy the floodplain, flood control structures are usually built to protect them. The reliance upon massive engineering works, like dams and levees, minimises the damage from frequent floods, but may contribute to deaths and greater destruction from less frequent major floods. Extensive flood protection works inspire an illusion of safety that may promote dense occupation of flood-prone areas. The stage is then set for enormous loss of life and property when these flood protection works fail or are overtopped or inundated by extremely heavy rains. A 1972 flood in Rapid City, South Dakota, killed 237 people and injured 3,057 when flood waters overflowed the storage reservoir and breached the dam upstream of the city. Many residents, confident of the dam's ability to protect them, stayed in their homes despite warnings to evacuate. The river rose fourteen feet in four hours and as much as 3.5 feet during a single fifteen minute period. The flood devastated 1,335 homes and demolished 5,000 automobiles. Of the estimated $160 million in property damage, less than $300,000 was insured.

Cities are not at equal risk to floods. A city's regional climate and seasonal pattern of rainfall, the amount of floodplain within the city, and the extent to which the floodplain is developed all contribute to the relative degree of flood hazard. Coastal cities in the eastern United States lie in the path of hurricanes and are prone to flooding from a combination of heavy rainfall and surging flood tides. Flood hazards on the West Coast of the United States are increased by the added threat of earthquake-generated tsunamis. Cities in semiarid and and climates may also have flood hazards; their shallow, wide floodplains, relatively dry much of the year, may be deceptive. James Micheners described the South Platte River that flows through Denver as a "sad, bewildered nothing of a river . . . a sand bottom, a wandering afterthought, a useless irritation, a frustration, and when you've said all that, it suddenly rises up, spreads out to a mile wide, engulfs your crops and lays waste your farms."

Most of the year, the South Platte consists of a shallow trickle engulfed in a wide, flat, sandy floodplain, but heavy seasonal rains convert the river into a raging torrent. In June 1965, fourteen inches of rain fell over parts of Denver within a few hours. Flood waters rose quickly, overflowed the banks and slammed debris against bridges, forming dams so that the flood surged around them into the adjacent city. When the storm had passed, most of Denver's bridges were destroyed, and highways and buildings buried in tons of silt. The flood was the worst disaster in Denver's history, taking twelve lives and costing $300 million in damages.

The extent to which the floodplain is constricted and built upon can aggravate the city's natural flood hazard. The amount of floodplain a city contains and the proportion of that area that is developed varies from city to city. Eighty-one percent of Monroe, Louisiana, and 40 percent of Charleston, South Carolina, lie within the floodplain, while floodplain comprises only 2.4 percent of Spokane, Washington. The design of a city's storm drainage system can also aggravate or alleviate flood hazard. The faster stormwater reaches streams and rivers, the more floods increase; the more stormwater is retarded, the more floods are reduced.

The effect of a storm drainage system is not limited to flood hazard; it can also increase water pollution and water use. Typically, the storm drainage system aggravates pollution by delivering slugs of sewage and runoff after storms and by decreasing stream flow between storms so that discharges from industry and treatment plants are undiluted. Cities that draw their water supply from urban rivers must then contend with vacillating flows and increased contamination. When sewage and stormwater systems are combined, as they are in many older cities, the surge of stormwater following a rain frequently overwhelms the capacity of sewage treatment plants, so that both rainwater and untreated sewage dump directly into water bodies. Since the ground, sealed by pavement and drained by pipes, absorbs little water, the amount of water stored in the ground, from which plants obtain their supply, is reduced. The lowered groundwater is insufficient to maintain stream levels between storms and sustain plants during dry spells.

Poisoned Water

The disgusting odor and appearance of water in the wells and rivers of dense cities has been a source of concern for centuries. Although, in the fourth century B.C., Hippocrates had warned that polluted water posed a serious health hazard, it was not until 1854, when John Snow, a London physician, traced the source of a cholera outbreak to polluted water from a single well, that the link between water and disease was definitively established. In thirteenth-century London, both the Crown and the City attempted repeatedly and ineffectively to halt pollution of the Thames, but the river continued to be an open sewer. The Thames was a grossly polluted river in 1855, when Michael Faraday complained in a letter to the *Times* that "the whole of the river was an opaque pale brown fluid . . . near the bridges the feculence rolled up in clouds so dense that they were visible at the surface." The following year, 1856, was the "Year of the Stink," and sheets soaked with disinfectant were hung in Parliament to combat the stench of the river. A century later, in the 1950s, the Thames was still so polluted that it was virtually fishless for a forty-three mile stretch in the proximity of London.

Recurrent epidemics swept nineteenth-century European and North American cities with terrifying frequency. Cholera epidemics hit London in successive outbreaks: in 1832, 1848, 1849, 1853, and 1854. Cholera killed 3,500 New Yorkers between June and October of 1832; during the height of the epidemic, 100,000 people, approximately half the population, fled New York. Pathogenic organisms bacteria, protozoa, worms, viruses, and fungi are responsible for outbreaks of waterborne diseases. The diseases they cause range from potentially deadly bacterial infections, like cholera and typhoid fever, to intestinal parasites and skin rashes. Most pathogens enter surface water via human and animal feces. Inadequately treated sanitary sewage and urban runoff account for nearly all water contamination by pathogens. As municipal sewage treatment improves, the pathogens present in urban runoff assume a new, until recently unrecognised, importance. Urban runoff has the bacterial contamination of dilute sewage and often exceeds concentrations considered safe for water sports by two to four orders of magnitude. The city's dog population contributes an enormous load of untreated sewage to urban runoff. The water near storm and sanitary sewer outfalls exhibits the highest concentration of pathogens, and is most contaminated immediately after a storm.

The specter of waterborne epidemic disease which haunted cities of the past has been laid to rest in the twentieth century by sewage treatment and the chlorination of public water supplies, but new poisons now threaten drinking water. The impact of cholera and typhoid fever was felt overnight, and their cause, once recognised, was swiftly eradicated. In contrast, the effects of the new poisons are gradual and cumulative. The diseases they generate and the genetic change they precipitate will not become fully evident for years, at which point they may not be readily removable from the environment. To complicate matters further, many of these pollutants have synergistic effects which increase their toxicity; some combine with chlorine to produce new, toxic compounds.

The Environmental Protection Agency has identified 129 "priority toxic pollutants," including heavy metals, pesticides, and organic toxicants. Many are poisonous even in extremely small concentrations, and in low doses over a long period of time can cause neurological damage, cancer, miscarriages, and birth defects. Extremely low, but harmful, concentrations of heavy metals, pesticides, and organic chemicals are often difficult to detect and to remove from water." The existence of so many toxicants also complicates both their measurement and impact. Toxic chemicals are a by-product of modern industrial processes, agricultural practices, and fuel consumption. Toxic pollutants enter streams, rivers, and lakes in industrial discharges, in urban stormwater runoff, and in the fallout of urban dust; they leach into groundwater from sanitary landfills, toxic waste disposal sites, and chemical spills. A 1977 study of surface water quality by the U.S. Environmental Protection Agency demonstrated that heavy metals and synthetic organic pollutants are a significant and widespread problem in water near industrial areas. As industry processes waste more effectively, urban runoff is emerging as a major source of toxic pollutants. Every heavy rainfall sweeps the dirt and debris of the city streets into storm sewers, and with it heavy metals and other toxic materials, oil, and grease.

Turbidity and warmer temperatures, the increase of nutrient salts, and the loss of dissolved oxygen degrade the water quality in urban rivers, streams, and lakes. These factors have less dramatic effects on human health than do pathogens and toxicants, but drastically affect aquatic life and may produce smelly, dirty water with a strange taste. Urban rivers are turbid; the suspended sediment in urban runoff is the major cause of turbidity, but solids from domestic sewage and industrial discharges are also factors. When nutrients like nitrogen and phosphorus reach rivers and lakes in large quantities, they trigger a prolific bloom of algae that chokes waterways with living and decaying plants. As plants decay, they consume dissolved oxygen and produce an unpleasant smell. Fish and many aquatic plants require oxygen, and the most sensitive species die as dissolved oxygen decreases. Lack of oxygen was the major cause of the lack of life in the Thames River in the 1950s. Nutrients enter surface water in sewage and urban runoff containing animal feces and fertilizers.

The character and severity of the water pollution problem varies from city to city. A city's major industries, the degree and type of air pollution, the nature of its sewage treatment and storm drainage systems, and the existence of industry, agriculture, or other cities upstream all determine which water pollutants are a problem. The most unfortunate cities are those, like New Orleans, which are located near the mouths of major rivers, downstream of millions of pollutant sources. The fate of New Orleans' water supply is beyond the city's control.

In 1977, the Council on Environmental Quality studied EPA records of water quality in 159 cities. The average concentration of bacteria exceeded levels considered safe for drinking water in onequarter of the samples. In Philadelphia, Charlotte, Roanoke, Omaha, and Denver, bacteria exceeded safe levels over 90 percent of the time." Cities which draw their water from lakes and rivers polluted with such high levels of bacteria are caught in an increasingly difficult dilemma. On the one hand, water must be treated with chlorine to prevent the spread of epidemic disease; on the other hand, chlorine combines with some organic pollutants to produce new carcinogenic compounds. Mercury is a problem in all of the twelve major United States river basins sampled by the Environmental Protection Agency in 1977; concentrations exceeded water quality criteria in more than three-fourths of the sample stations, with median values ranging from eight to forty times the standards set by the EPA for the protection of aquatic life. Concentrations of cadmium and selenium also exceeded the proposed EPA criteria for water quality in at least 10 percent of all the samples.

A city's regional climate and precipitation patterns, its underlying geological conditions, the character of water circulations in its rivers, streams, lakes, ponds, and marshes, the types of land uses occupying flood-prone areas, the pattern of its sewage system, and its urban form all these factors influence where, when, and how water pollutants are concentrated or diluted. Lakes may be more susceptible to pollution than rivers. Water in a river moves steadily toward the mouth; water circulation in lakes is more complex. Circulation time, the time it takes the water in a lake to be completely replaced, varies with the size of the lake's drainage basin, the amount of rainfall it receives, and the depth and surface area of the lake. Circulation time determines how susceptible the lake or pond is to pollution. The longer the circulation time, the more sensitive the lake to contamination, and the more difficult its recovery. Urban harbors and marinas, whether on lakes or rivers, are protected from currents and wave action and have reduced water circulation; therefore, like small lakes and ponds, they are highly sensitive to pollution. Trash and other pollutants accumulate in slips and canals that receive little flushing.

Although lakes and rivers are generally more contaminated than groundwater, they exhibit pollution more quickly and respond to improvement more rapidly. The quality of groundwater is less easily monitored than surface water. Pollution may go undetected until it reaches a well, at which point the source of contamination may be difficult to locate. Water moves very slowly through the ground, and abandonment may be the only alternative when a well becomes contaminated. Leaks from sewers, disposal of toxic industrial wastes, leaching from sanitary landfills, salt from highway de-icing, fertilizers and pesticides, leaks from chemical storage tanks, and the intrusion of sea water or saline groundwater are increasingly polluting groundwater. The pollution of groundwater by hazardous waste now threatens the public water supplies of Tampa, Florida, and Atlantic City, New Jersey, a reservoir in King of Prussia, Pennsylvania, which supplies drinking water to 800,000 people, and the water supplies of countless other communities, many of them as yet undocumented.

Dwindling Water Supplies

Without water, a city cannot survive. Disputes over water rights were among the most bitter and violent struggles in the history of the American West. Today, cities separated by a third of a continent, Denver and Los Angeles, dispute the use of the same Rocky Mountain water. Within the next decade, many cities will face a major water crisis.

The combination of contamination and lowered groundwater has always threatened city water supplies. Privies and graveyards befouled wells, and garbage and sewage polluted rivers and lakes. Until the twentieth century, Chicago dumped its sewage into and drew its water from Lake Michigan. In 1891, typhoid fever took 2,000 lives, a death rate of 173 out of every 100,000 citizens. Chicago cut this death rate by almost 90 percent by diverting its sewage away from Lake Michigan. The construction of the Chicago Drainage Canal in 1900 reversed the flow of the Chicago River, so that sewage flowed to the Mississippi River. This proved a fine solution for Chicago, but created new problems for other cities downstream on the Des Plaines, Illinois, and Mississippi rivers. Other cities, like Boston and New York, had earlier opted to abandon local wells and to import water from distant reservoirs.

The alteration of the city's hydrology by pavement and sewers and their effect on both water availability and water quality had been recognised well before the twentieth century. Benjamin Franklin left a legacy to the city of Philadelphia, recommending that it be used to secure a public water supply. His will, read in Philadelphia in 1790, stated: And having considered that the covering of the ground-plot of the city with buildings and pavements, which carry off most of the rain, and prevents its soaking into the Earth and renewing and purifying the Springs, whence the water of wells must gradually grow worse, and in turn be unfit for use, as I find has happened in all old cities, I recommend that at the end of the first hundred years, if not done before, the corporation of the city Employ a part of the hundred thousand pounds in bringing by pipes, the water of the Wissahickon Creek into town, so as to supply the inhabitants . . .

Figure : Water supplies are dwindling.

Franklin's prophecy regarding the pollution of urban wells was borne out in Brooklyn, New York. From its initial settlement until 1947, Brooklyn depended on well water. To avoid contamination by surface cesspools, wells were drilled to ever-increasing depths. By 1936, following the installation of sewers and pavement of streets, accompanied by increased pumping, the water table dropped more than thirty-five feet below sea level. The saltwater contamination that resulted led to the abandonment of virtually all the wells by 1947. With pumping halted, the water table gradually rose again, flooding basements and subway tunnels constructed when the water table was lower and causing hundreds of thousands of dollars in damage. Brooklyn, like many suburban communities whose wells have become contaminated, tied into the larger metropolitan water supply system, further increasing the demand for distant water sources. The problem repeats itself in the remainder of modern Long Island, completely dependent upon groundwater, whose wells are continually threatened by contamination and salt-water intrusion.

Approximately three-quarters of all American cities obtain their water supplies from groundwater and three of the thirty-five largest rely on local groundwater alone Miami, San Antonio, and Memphis. Of the remaining thirty-two, fifteen tap either the Great Lakes or water from major rivers, and twelve garner water from a combination of sources, often importing water from great distances. Each city not only competes with other cities for water but also with local industries that obtain their own water. Supply has never kept pace with demand. Cities must constantly search further and further afield to appropriate water. Only cities which draw from a vast, uncontaminated reservoir of groundwater or from a large, freshwater lake or river are exceptions. Much of New York City's water comes from the Catskill Mountains over one hundred miles away; Boston's water from the Quabbin Valley in central Massachusetts sixty-five miles away; and Los Angeles diverts some of its water from the Colorado River, with its source on the west slope of the Rocky Mountains over six hundred miles away. As growing, dispersed suburban and rural settlements obscure the boundaries between cities, and as the central city loses political power, cities find it more difficult to appropriate distant water supplies.

At the same time urban water supplies are threatened by contamination and depletion, water is squandered. Americans have long used more water per capita than Europeans. The average per capita use in London, Berlin, and seven other European cities was only 39 gallons per day before World War II. During that same period, the average daily consumption in Ten American Cities was 155 gallons, or nearly four times that amount. By 1975, per capita water use in the United States had reached 168 gallons per day. The average American uses 20 to 80 gallons per day at home. It takes approximately 6 gallons to flush an average toilet, 20 to 40 gallons for a bath, and 20 to 30 gallons to run a washing machine. A leaky faucet dripping one drip per second wastes 4 gallons per day. Watering a garden of 8,000 square feet requires 80 gallons a day in a humid climate and 500 gallons per day in an and climate.

Uncontaminated fresh water is a diminishing resource. Using drinking water to flush toilets and water lawns is a scandalous waste. Increased industrial demand for water, the invention of domestic appliances like washing machines, and the popularity of a pastoral landscape which requires extensive irrigation, have all contributed to spiraling water use. On the average, domestic use of water accounts for approximately one-third of the water withdrawn from municipal water supplies. Industry utilises water mainly for cooling and accounts for over a third of the water demand, on the average, but may represent a much greater proportion in some cities. Commercial and public use of water and water lost through leaks in underground pipes account for the remainder. The amount of water lost through leaks is probably equal to the sum of all public water use: for fire fighting, street cleaning, park irrigation, and water for public buildings, swimming pools, and fountains.

Together, dwindling, poisoned water supplies and flooding represent the most significant threats to health and safety of city residents. Water comprises approximately three-quarters of our body. No other resource affects the health of every citizen so intimately and thoroughly, yet cities continue to operate, as they have throughout history, with marginal water systems.

Cities respond to each water crisis with narrow solutions which address immediate needs at minimum cost, but ignore the need to promote water conservation and to overhaul overtaxed and outdated collection, storage, and distribution systems. Even as the city thirsts, rainfall is not permitted to enter the ground, but is quickly diverted by the storm drainage system. Parks are built with more pavement and fewer trees, permitting less water to infiltrate the ground. Storm sewers drain the rainfall from parks, and water sprinklers irrigate plants. A water-demanding aesthetic of trees and lawns proliferates in the parks of cities in semiarid and and climates, further straining the paltry water supply and polluting it with fertilizers, pesticides, and herbicides.

Toxic heavy metals and organic chemicals represent the greatest waterborne threat to health since the epidemics of infectious disease in the eighteenth and nineteenth centuries. Industry and waste disposal sites are located on aquifer recharge areas, and contaminants seep into groundwater. Storm sewers deliver their complement of toxicants to surface water. As new development locates in headwaters, and houses and businesses crowd and constrict the floodplain, the magnitude of flooding and the damages it inflicts increase. Cities must manage their water resources more wisely. At stake is survival itself.

Improving Water Quality By Reducing Waste Discharges

Factors Reducing Wastes Generation

The generation of industrial wastes per unit of raw product processed or per unit of final product is a function primarily of the type of raw materials used in the production process, the technology of the production process, the product mix, and sometimes the extent of in-plant water recirculation.

Thus there is a wide range in the wastes generated per unit by different industries, and also by different plants within a single industry. To illustrate, a German study reported ratios of high to low waste discharge (in pounds of BOD per unit of product or raw product processed) within several industries as follows: about 25 for paper mills, 2.5 in malt factories, 6 in starch factories, 50 in pork slaughterhouses, about 10 in tanneries, and about 20 in textile factories. Similar variations occur in U.S. industries.

Shifting to a different grade or type of raw product is one way of reducing waste loads. The sulfur content of crude oil has an important bearing on a petroleum refinery's waste. In pulp and paper, the type of wood used is significant. In fruit and vegetable canning, the size, shape, per cent of solids, and resistance to damage and bruising in harvesting and processing the raw product affect the waste load per ton of raw product processed.

Changes in production processes, however, have the major impact on wastes generation, although it should be emphasized that technological changes have not been, and are not usually, instituted because of water quality problems. In fact, most changes in production technology have been stimulated by factors unrelated to water problems, and have been developed without explicit consideration for their effects on water quality.

On the other hand, where various stimuli to management, such as sewer charges, have been used, they have often resulted in process modifications. As more systematic means are developed to bring to bear on industry the broader social costs associated with the discharge of waste materials into the environment, the generation and control of wastes will receive more prominent consideration in process design, and perhaps even in product development.

Studies of several industries make clear that even comparatively small design changes can have a substantial effect on the generation of wastes.

Table: *Production of Wood Pulp in the United States, by Process and Year (per cent of total production)*

Process	***1940***	***1945***	***1950***	***1955***	***1960***	***1965****
Mechanical	18.2	18.0	14.9	13.2	13.0	11.8
Unbleached sulfite	11.1	8.0	5.0	3.1	2.1	1.5
Bleached sulfite	18.0	15.2	14.2	13.9	12.6	11.4
Unbleached sulfate	35.3	35.6	38.4	37.0	34.3	35.4
Bleached sulfate	6.5	8.4	12.1	17.5	23.3	26.3
Semi-chemical			4.6	6.8	7.9	8.7
Soda	5.9	4.2	3.5	2.1	1.7	0.7
All other	4.9	10.6	7.2	6.4	5.2	4.4
Total	99.9	100.0	99.9	100.0	100.1	100.2

Source: American Paper and Pulp Association, *Statistics of Paper-1964*, p. 10, and American Paper Institute, *Statistics of Paper, 1966 Supplement, 1966*, p. 2.

* = Preliminary.

Includes special alpha and dissolving grades.

Includes semi-bleached sulfate.

Reported in "all other" for 1940 and 1945.

Total production increased from 8.96 million short tons in 1940 to 33.3 million short tons in 1965.

In the pulp and paper industry the proportions of the total output produced by different processes have shifted significantly over the past two decades. The largest and most significant shift has been from the sulfite process to the sulfate process, which, by making the recovery of chemicals for internal reuse economic, can reduce the waste load in terms of pounds of BOD per ton of product to about 5 or 10 per cent of the previous level. The development of save-alls to recover waste fibers which formerly were simply discharged into water courses was stimulated by the increase in the value of fiber. Their use has resulted in a significant reduction in terms of pounds of both BOD and suspended solids per ton. Both black liquor recovery systems and save-alls were stimulated originally by the economic incentive to save materials useful in the production process, rather than by waste problems.

Other possibilities for process changes which will reduce wastes generated appear to be on the horizon. One of these is a "dry" papermaking process that will use considerably less water as a defibrating and distributing agent in the paper-making process. A particularly difficult waste problem in the steel industry for many years has been disposal of the "waste pickle liquor" generated when steel is pickled with sulfuric acid. Attempts to develop an economically feasible sulfuric acid recovery process have met with little success under U.S. conditions.

Another solution is to shift to hydrochloric acid (HCI) in the pickling process itself. A recent report indicates that the acid losses in the effluent are reduced by this process change from about 13 pounds of sulfuric acid per ton of steel pickled to about 0.2 pound of hydrochloric acid per ton of steel. Hydrochloric acid offers other advantages as well a high-quality product (in terms of brightness of surface) and faster production. In the beet sugar industry the waste load generated in pounds of BOD per ton of beets processed has been reduced greatly in the last two decades by comparatively simple and economical changes in processes.

Table: *Wastes Generated in Beet Sugar Plants, by Process1 (pounds of BOD per ton of beets processed)*

Operation	***Potential BOD generation***	***After process changes***	***After recirculation associated with process changes***
Flume water	4.5	4.5	4.5
Screen water	2.52	2.53	0
Press water	2.6	2.63	0
Silo drainage	12.3	04	0
Lime cake slurry	6.5	6.5	6.5
Condenser water	0.7	0.7	0.7
Steffens waste	10.4	05	0
Total	39.5	16.8	11.7

Source: G. 0. G. Löf and A. V. Kneese, *The Economics of Water Utilisation in the Beet Sugar Industry* (Resources for the Future, Inc., 1968).

Based on typical 2,700 tons per plant day.

Average of BOD waste generation in batch diffusers and continuous diffusers.

Assuming conversion to continuous diffusers but without recirculation of screen and press water.

Assuming pulp dryer installation.

Assuming discontinuance of Steffens processing or the complete use of Steffens waste in by-product manufacture.

The substitution of drying of beet pulp for storage of wet beet pulp in silos and the use of Steffens waste for the production of by-products, or the complete elimination of Steffens processing can reduce BOD generation by about 60 per cent. Converting to pulp drying has another advantage if there is a market nearby, because dry pulp brings a greater return than wet pulp. The other process change, i.e., a shift from cell-type to continuous diffusers, is integrally related to recirculation of screen and press water. This change reduces the BOD generated by about 10 per cent.

Specific cases, as well as general industry examples, can be cited. For example, one chemical plant over a fifteen-year period doubled the dollar value of production and at the same time reduced the BOD and phenolic waste loads per day by over 90 per cent. The waste reductions were accomplished primarily by process modification, in many cases by simple changes such as increasing the temperature at which a reaction was taking place and changing a polymerization catalyst.

In the production of potato flakes, the processing of raw potatoes by steam peeling results in a BOD waste with a population equivalent of about 200 per ton in contrast to about 420 per ton using caustic peeling. Where cooling is a major component of an industrial process, changes in cooling methods are changes in production processes. The adoption of air cooling, as in some chemical plants and petroleum refineries, will reduce the thermal waste load generated in the production process.

It is interesting to note that air cooling may be the rational economic choice even where water is relatively inexpensive in terms of both supply and waste disposal problems. The cost associated with circulating a large quantity of air must be compared with the costs of high head pumps, pumping energy, and pipelines required to circulate water, plus the costs associated with water intake and wastewater disposal. There is obviously no waste to be disposed of from air cooling units. In addition to the lower investment costs with air cooling, there are operating savings from less equipment maintenance and improved continuity of operation.

In many cases it is difficult to separate changes in production processes from changes in products. Change in consumer demand over time is one stimulus for development of new production processes. Several examples illustrate the nature of product changes, with their implications for changes in wastes generation. In petroleum refining the product mix has shifted significantly over time toward the production of more gasoline per barrel of crude throughput. Continued changes in product mix are in prospect. For example, by 1985 the demand for jet-fuel will require about 16 per cent of all crude runs in the United States, in contrast to only about 3 per cent in 1965. This will require additional use of hydro cracking with consequent changes in waste generation per barrel.

In pulp and paper manufacture, there has been a steady shift toward a wider variety of products, such as colored paper and coated products of various types. Many of the coatings and sizings which have been developed to provide particular characteristics in the final products make reuse of the waste fiber more difficult or impossible, with a consequent increase in the waste load generated per ton of product. There also has been an increase in the proportion of pulp output which is bleached, from about 25 per cent in 1940 to almost 40 per cent in 1965. Bleaching increases the waste load generated in pounds of BOD per ton by about 25 per cent.

An example of a significant change in waste load through product change is the shift from hard detergents to soft detergents which are readily degraded by the biota in natural watercourses and in waste treatment plants. The change took place in Germany in October, 1964, and in the United States in the summer of 1965. Even so, soft detergents are still "under fire" because of their phosphate content, which contributes to algae growth in streams and lakes, and research is now oriented toward finding a substitute for phosphate.

Process and product changes have obviously had different effects on wastes generation. However, the dominant tendency has been toward a reduction in wastes generated per unit of raw or finished product. Two points are important:

(1) most of the changes in production processes and product mixes have occurred independently of problems of water quality management (and of air and solid wastes management);

(2) there are usually several ways at least of producing a given product mix, so that there are possibilities for modifications of production processes to reduce the wastes generated per unit.

One other type of in-plant change that can reduce waste generation is in-plant recirculation of water. As the extent of water recirculation in a plant increases, a decrease may occur in the waste load generated per unit, e.g., pounds of BOD and/or suspended solids per ton. This has occurred in the beet sugar industry and in the canning of some fruits. In the latter, where flume water is recirculated, less sugar is dissolved from the fruit in processing, thereby decreasing the BOD load per ton.

The separation of the effect of water recirculation *per se* on waste generation from the effect of changes in production processes and from materials recovery is not clearcut. In-plant recirculation of water may require modifications of production processes. Conversely, some production processes enable recirculation of water, as noted in the previous description of process changes in the manufacture of beet sugar. It is clear, however, that water recirculation in some cases has significantly reduced waste generation per unit.

Methods for Reducing Wastes after Generation

Figure: *Big brand reduce waste.*

Materials recovery and by-product production are two methods for reducing wastes after generation. They form a logical pair, because the types of processes are often very similar. We differentiate between them on the basis of the final destination of their outputs. In materials recovery the output is reused within the same production unit as an input to the production process. By-product production yields consumption goods or intermediate goods used in other production processes. Examples of materials recovery are to be found in almost all, if not all, of the heavy water-using or water-polluting industries. Only a few will be mentioned here for illustration. The recovery of chemicals in black liquor recovery systems in the production of sulfate pulp and the utilisation of save-alls in paper production have been mentioned previously. In the manufacture of synthetic phenol by the sulfonation process, liquid wastes have been essentially eliminated by process engineering, and the value of recovered materials is reported to exceed recovery costs.

In steel production, sedimentation to eliminate mill scale from waste discharges has resulted in the recovery of iron that can be used in the production process. Plating compounds, a potentially troublesome waste load, can be recovered from rinse water by evapouration. This procedure concentrates the rinse water sufficiently for reuse in the plating bath, which not only conserves valuable plating compounds but also significantly reduces waste and thereby the outlay for neutralizing chemicals. One plant anmure. Waste liquor from citrus peel processing, normally a waste disposal headache, is concentrated to syrup and returned to the process. This procedure saves fuel as well as reducing the waste.

A number of industries have achieved substantial reductions in their final waste loads through by-product production. In beet sugar production, the recovery of monosodium glutamate and potash from the Steffens waste has significantly reduced the BOD generated.

In canning apples, the utilisation of the "waste" segments of the apple for vinegar production has reduced the waste load per ton. Cottage cheese whey has been converted into protein food supplements, thereby eliminating a difficult waste problem. Some of the by-products from waste materials in food processing are quite imaginative. For example, bran waste from milling has been used to make inexpensive sets of plates, casseroles, and trays, and coffee grounds have been used to produce discs and shelves.

As indicated, many alternatives are available for the reduction of wastes after generation by materials recovery and by-product production. In many cases these methods result in net profits to the industrial operation, even without counting the broader social costs avoided by the reduction or elimination of wastes discharged to watercourses. After economic processes of materials recovery, by-product production, and effluent reuse have been exhausted, there are often residual waste loads which may have to be reduced or otherwise handled prior to final discharge. Some common methods for treating industrial wastes, also applicable to municipal wastes, are listed below:

Screening	Neutralization	Incineration
Flocculation	Chemical oxidation	Biological filtration
Chemical coagulation	Chemical reduction	Activated sludge
Flotation	Wet oxidation	Anaerobic digestion
Sedimentation	Fermentation	Stabilisation lagoons
Centrifuging	Emulsion breaking	Spray irrigation
Filtration	Evapouration	Disinfection
Stripping	Distillation	

In addition, the waste loads from some industrial operations may contain substances requiring special treatment, Such as colours, odors, special chemicals, and toxic materials.

The waste treatment methods listed above might be characterised broadly as physical, chemical, and biological processes. They have greater or lesser applicability to the two major types of wastes, nondegradable and degradable. With respect to the former, suspended and dissolved solids are the major wastes. Suspended solids can be removed by sedimentation, with or without the aid of flocculants such as polyelectrolytes, by filtration through various kinds of screens hind filters, and by centrifuging. Dissolved solids can be removed, to any desired degree, by one or more of the processes of distillation, ion exchange, electrodialysis, and reverse osmosis. It is possible to obtain completely pure water from waste discharges, but there will still be some sort of residual waste, either a concentrated brine or a semi-solid sludge. Ultimate disposal still remains a problem.

The degradable wastes of primary concern ire organic wastes, bacteria, and thermal discharges. Organic wastes are removed by physical processes, such as sedimentation, to some degree, and by chemical and biological processes, such as chemical oxidation and aerobic digestion, respectively. Bacteria are at least partially removed by biological processes and by disinfection. Thermal loads are removed by one or more types of air or water cooling systems.

Conventional treatment of domestic wastes consists basically of physical and biological processes, and is a comparatively standardised process. It can greatly reduce BOD, but will not completely, eliminate it, and is frequently designed to improve bacteriological quality and essentially eliminate suspended solids. Ordinarily the treatment of degradable organic wastes either domestic or industrial begins with the removal of the larger suspended solids by screening and grit chambers and the more finely divided suspended solids by sedimentation. New practices, such as the addition of polyelectrolytes have done much to improve the efficiency of the sedimentation process. This sedimentation stage, known as primary treatment, results in a wet, difficult-to-handle sludge, which is usually digested in heated anaerobic tanks before final disposal. In some cases the sludge is burned, usually after some degree of drying, or concentrated with centrifuges, prior to disposal. When waste treatment plants are operated adequately, primary treatment of an effluent containing an organic waste, with sludge disposal, can reduce the first stage BOD by 35 to 40 per cent.

What is generally termed a secondary stage of treatment is biological in character and essentially controls and accelerates the oxidation processes which occur in natural waters. Two major techniques are presently used, with various modifications. One is the trickling filter, in which a biological film is grown on rocks or some type of plastic medium, and the waste is applied intermittently, allowing air contact with at least the surface of the film. The other technique is the activated sludge process, in which air or oxygen is forced into a tank containing a mix of waste and actively feeding biota. Part of the settled sludge containing the biota is recirculated and mixed with the entering waste in the aeration chamber.

Each of these methods has certain advantages. Trickling filters require relatively little attention during operation, and the biological growths upon which the effectiveness of the filters depends are somewhat less susceptible to toxic substances than the activated sludge process. An activated sludge plant is somewhat more flexible in operation and design; the major variables are the proportion of sludge recirculated, the length of aeration time, and the amount of air or oxygen introduced. Rapid turnover of waste and sludge generally saves space and power, but requires well-trained personnel and continuous monitoring of plant performance. Treatment in all activated sludge plants can be made more effective by increasing the quantity of air passed through the sludge tank, although there is a definite limit to this procedure. Costs in activated sludge and other types of treatment plants increase rapidly as the removal of BOD is pushed above 95 per cent.

Small package plants, called extended aeration plants, have been used with considerable success for handling relatively small volumes of wastes, usually from strictly domestic sources. These plants are easy to operate, the sludge withdrawals are small enough in volume to be discharged with the final effluent, and much more of the solid matter is in mineralised form. When secondary treatment is undertaken, the supernatant liquor from the sludge digestor, or other liquid resulting from the handling of sludge, is routed through the secondary treatment process. Secondary treatment in turn gives rise to some additional sludge which is routed into the digestors. Primary and secondary treatment combined usually can reduce BOD by between 80 and 95 per cent. The procedures outlined above, plus disinfection of the final effluent in some instances, have been the essential elements of the primary-secondary waste treatment sequence for more than fifty years.

Although the conventional treatment processes are effective in reducing BOD, they have been much less effective in removing plant nutrients, nitrogen and phosphorus. During secondary treatment the nitrogen content of the organic waste is successively converted to nitrites and nitrates. Over-all, the volume of potential plant nutrients is reduced modestly during secondary treatment, but the content of nitrogen in nitrate form of course increases. Within limits, the extent to which nitrification occurs in the treatment process rather than in the receiving water can be controlled.

Other treatment methods that have become increasingly common and that achieve waste reduction comparable to secondary treatment are oxidation ponds (or stabilisation lagoons) and spray irrigation systems. Both methods are used in the food processing and pulp and paper industries. Ponds are used by the petroleum refining industry, by some large metropolitan areas (Melbourne, Australia, for example), and increasingly by smaller communities. Ponds, lagoons, and spray irrigation are used alone as complete treatment processes or as secondary or tertiary treatment in combination with other treatment processes. Ponds and spray irrigation systems may also be used in series, as in a number of installations in the food processing industry.

Stabilisation lagoon or oxidation pond systems in industrial applications may range from a single large pond to a series of as many as eight or nine ponds. Pond capacity may be sufficient to store the entire waste effluent from a plant with a seasonal production schedule, as in some beet sugar and canning installations, or during the low streamflow period for plants with continuous production schedules. Alternatively, the pond system may be operated to produce a continuous outflow, as occurs in some installations in the canning, petroleum refining, and pulp and paper industries. A series of ponds may also have aeration devices of one kind or another between ponds and or floating or fixed aeration devices on or in one or more of the ponds themselves.

Algae play an important role in ponds and lagoons, as they furnish a substantial share of the oxygen requirements of aerobic bacteria, and they utilise the gases from the anaerobic BOD removal in the bottom sludge banks of such ponds and reduce or prevent their escape to the atmosphere. Such ponds and lagoons, when properly designed and loaded, are capable of stablizing oxygen-demanding wastes to a high degree. In the Ruhr industrial area of West Germany, where comparatively large, shallow impoundments have been constructed in several rivers to serve essentially as large oxidation ponds in the streams themselves, the impoundments are sometimes specially designed to handle the specific types of wastes they receive. One, for example, neutralizes acids and precipitates organic materials simultaneously.

One effect of the use of ponds, lagoons, and spray irrigation systems (and underground disposal) is the increase in the consumptive use of water per unit of raw product or final product, compared with conventional treatment. Data from the petroleum refining industry show that the mean consumptive use was nearly 50 per cent greater for refineries using ponds and underground disposal than for those using conventional treatment. Although the increase was small in absolute terms about 14 gallons per barrel of crude throughput it could be significant during periods of low streamflow, as in the summer months when the consumptive use in waste treatment by these methods and demands for water are both at a maximum.

It is important to note that both the oxidation pond or stabilisation lagoon and the extended iteration treatment methods differ from standard waste treatment in that their effluents contain virtually all of the plant nutrients originally in the waste discharge. In the ponds, they are embodied in algae; in the latter, in mineral form. Thus the effluents from these types of treatment processes fertilize receiving waters more than effluents from conventional treatment plants do. Considerable attention is being devoted to developing methods for removal of plant nutrients from final effluents. Some recent results suggest that progress is being made on this problem. If fish or algae were harvested from oxidation ponds, the process of biological treatment would be accelerated and a major portion of the plant nutrients, as well as some other chemicals, would he extracted from the waste water. Chemical treatment includes a variety of processes ranging from chemical oxidation of substances in waste discharges to neutralization of acid or alkaline wastes. Most types of chemical treatment involve low capital costs and high operating costs; this makes them particularly useful for specialised treatment problems, for final finishing of a sequence of waste treatment, and for infrequent application during critical periods of water quality. Some laboratory experiments with chemical treatment have shown a reduction of 60 to 85 per cent in BOD without biological treatment. Disinfection, by use of chlorine or ozone, can also be considered a type of chemical treatment.

One final point should be made concerning waste treatment. The increasingly stringent quality standards of final effluents make the economics of in-plant water recirculation more favourable. In some cases, water may be suitable for reuse within a plant even as process water and yet not meet the quality standards for effluents.

The other major type of degradable wastes consists of thermal loads, which are of increasing importance. The most common method of reducing the heat load discharged is the installation of cooling towers. The magnitude of heat load reduction possible by this process is exemplified by the installation at the Philadelphia refinery of the Atlantic Refining Company, where the cooling tower program resulted in an 85 per cent reduction of heat load discharged to the river.

Whether or not a cooling tower is installed at a particular plant is a function of both the internal economics of over-all water utilisation in the plant and of the offsite costs resulting from the discharge of thermal loads. With respect to cooling towers, the economics of recirculation are such that if recirculation is adopted, it will be done essentially completely. As a result, little or no heat will be discharged, but there will be an increase in the discharge of dissolved solids, stemming from chemical treatment of the recirculated water and concentration from evapouration.

Cooling towers may also aid in reducing other waste loads. For example, in the Toledo refinery of Sun Oil Company, a cooling water stream containing phenols passes through a cooling tower which functions essentially as a trickling filter. Biological oxidation reduces the phenol content in the effluent by over 99 per cent. Other applications of the same procedure have been reported.

A final method for handling residual wastes is the reuse of a municipal or industrial effluent by a subsequent user or users, usually after one or more types of man-made or natural treatment. Sewage effluent is used for various purposes in industrial operations, particularly in the southwestern United States. The practice is less common in the East, although effluent from the Baltimore municipal treatment plant is used in the Sparrows Point plant of Bethlehem Steel, primarily for cooling water in steel production. Effluents from municipal waste treatment plants have long been used for irrigation. Among the many cities that use all or part of their sewage effluents for irrigation are Grand Canyon and Tucson in Arizona; Bakersfield, Fresno, San Bernardino, and San Francisco in California; and over 200 towns in Texas including Abilene, Kingsville, Lubbock, Midlands, San Angelo, and San Antonio.

Different reuse methods have varying effects on residual waste loads. Irrigation tends to remove virtually all of the residual wastes from the effluent, while industrial reuse may or may not reduce the residual waste load. Irrigation use, however, results in considerable net depletion of water, which is not necessarily unproductive. Artificial recharge of groundwater aquifers represents another method of effluent reuse. In this case, however, generally a larger number of intermediate processes are involved than in the cases mentioned previously.

Usually the effluent from a municipal waste treatment plant is inserted into a groundwater basin by means of injection wells or spreading grounds. In the latter case, the wastewater undergoes additional purification as it passes through the unsaturated zone. The purification process in the soil results from the action of soil bacteria, filtration, adsorption, and perhaps other processes. In other cases the waste effluent from the sewer system is conveyed directly to a wastewater reclamation plant, as at Whittier Narrows, California, where it is treated and then used for groundwater recharge. Another extensive system has been in operation for several years at Santee, California, where treated effluent is discharged into a series of ponds, the final ones having now been approved for water-contact sports. The final outflow is recharged to the groundwater basin. Caution is necessary in using wastewater from domestic and/or industrial operations for groundwater recharge. Even with treatment the minerals in the effluent before recharge remain, so that the total dissolved solids content is likely to increase (although there may be wide dispersion in the groundwater aquifer). Further, because there is likely to be some residual load even after treatment and the process of groundwater recharge, it is possible for groundwater contamination to occur. However, the well-publicised quality degradation of groundwater resources in certain areas, such as Long Island and Minneapolis-St. Paul, happened in situations where there was no centrally designed and operated wastewater treatment and recharge facility. Recent analysis suggests that groundwater recharge for water renovation may have much wider applicability and be more economical than has previously been appreciated.

An important point in connection with all measures of handling wastes is the existence of economies of scale. In materials recovery this is illustrated by the costs of the Blaw-Knox Ruthner sulfuric acid recovery process for handling waste pickle liquor. On the basis of 8.5 per cent free acid in the liquor, costs are about 3.5 cents per gallon when 30,000 gallons per day are processed and about 3 mills per gallon when 160,000 gallons per day are processed.

Similarly, significant economies of scale exist in waste treatment. For example, in a study of fruit and vegetable canneries, the costs (annual charge on capital investment plus operation and maintenance costs) for lagoons and/or spray irrigation systems were found to vary from about 2 cents per pound of BOD removed for operations removing more than 500,000 pounds of BOD in a canning season to 30 cents per pound for operations removing less than 100,000 pounds. Extensive analyses by the U.S. Public Health Service show similar economies of scale for municipal waste treatment plants. In densely developed areas the costs of collective facilities for conveying and treating municipal and industrial wastes can often be much less than the sum of the costs of separate facilities for each municipality and industrial plant.

Reducing Waste Discharges

The foregoing discussion has focused on the wide variety of methods by which the quality of receiving waters can be improved by reducing waste loads discharged to them. The various methods can be combined in a myriad of ways to meet the needs of a single industrial operation or urban area or a complex consisting of many operations and or areas. Considcer just waste treatment to reduce waste loads. Even in a single industry the combination of treatment processes applied in different plants varies widely, primarily as it function of the raw materials utilised, the production processes involved, the final products, and environmental constraints, including effluent controls. Consequently, there is a wide range in waste treatment costs in any given industry.

Figure: *Waste Discharge Charge System (WDCS)*

What can be done to reduce waste loads discharged to receiving waters is clearly exemplified by the performance of the beet sugar industry. The types of measures to reduce waste loads used by the beet sugar industry are likely to be integral components of optimal water quality management systems. The planning of such systems requires the development of relationships between the incremental quantity of wastes removed and the incremental cost of removal. .

Increasing the Assimilative Capacity of Receiving Waters

Because of the inverse relationship between the concentration of most water quality variables and water quantity, providing water for dilution by fuller use of available streamflow through flow regulation suggests itself as a possibility. Low streamflows often coincide with heavy concentrations of waste loads, as well as with high temperatures. Where the waste loads consist of oxygen-demanding wastes, the dissolved oxygen level of receiving waters is thereby depressed. Periods of low flow and potentially large reductions in water quality are the conditions for which effluent treatment plants and effluent redistribution systems are generally planned and designed. The assimilative capacity of tidal estuaries may also be significantly affected by variations in inflow; but the effects are more dispersed and difficult to analyse than in the case of flowing streams.

Dilution of wastes beyond that normally provided by natural streamflow during low-flow periods can be provided by modifying the pattern of streamflow, by modifying the time pattern and 'or locations of waste discharge, or by some combination of the two. The most common practices are to increase flows during low-flow periods by controlled releases from reservoir storage, and to modify the pattern of waste discharge by temporary storage of wastes. Flow can also be increased by withdrawing water from groundwater sources and releasing it into surface watercourses. An interesting variation of low-flow augmentation is the recently inaugurated Buffalo River Improvement Project, in which water is pumped from Lake Erie, used as cooling water on a oncethrough basis in five industrial operations, then discharged into the Buffalo River to maintain a minimum flow.

The effectiveness of augmenting flow depends on both the type of waste and the type of receiving water involved. With respect to organic wastes, for example, increasing assimilative capacity by flow augmentation is effective for streams, but not for lakes, and only to it limited extent for estuaries. Increased flows could actually reduce water quality in lakes and estuaries by carrying greater quantities of partially assimilated wastes into them. With respect to nutrients, increased flows would likely result only in a more rapid discharge of these substances into lakes and estuaries, because less time would be available for adsorption and sedimentation in stream channels. In contrast, with respect to nondegradable wastes such as chlorides, increasing, the flow into estuaries can result in a significant improvement in water quality. It appears difficult, if not impossible, to increase the assimilative capacity of groundwater bodies by augmenting inflows.

Flow augmentation may play a large role in future water resources management in the United States. Large-scale planning efforts of the U.S. Corps of Engineers in several basins in the East will include flow augmentation for water quality improvement its an important feature. The final report of the Corps on the Potomac River Basin recommended a large amount of reservoir storage for increasing low flows to improve water quality.

The effects of reservoir storage itself on the quality of water later released may be favourable or unfavourable. On the positive side, bacteriological quality tends to be stabilised, and summertime releases from reservoirs tend to be cooler than normal streamflows. On the other hand, some early hopes for effective dilution through releases from reservoirs were disappointed because of the failure to consider the impact of impoundment on the dissolved oxygen content of releases. Water from deeper parts of reservoirs is often virtually devoid of oxygen, because of the combined effect of BOD demand and reservoir stratification. When this condition is present as it is especially likely to be in areas experiencing hot summer seasons it takes larger releases from reservoirs to achieve it given effect on dissolved oxygen downstream, and the costs of achieving the desired water quality level are increased.

There are several ways of coping with the unfavourable effects of water storage in reservoirs. Multiple outlets can be installed as at the Oroville Dam storage unit on the Feather River in California so that water can be released from different levels in various combinations to achieve the desired water quality. If there is a power installation at a reservoir, air or oxygen can be introduced into water in the turbines. In many cases this can be done rather simply by using vacuum breakers already installed in the dam. Turbine aeration involves some power loss, and therefore is not costless.

Reservoirs can have significant effects on other elements of water quality such as suspended sediment, turbidity, and certain chemical constituents. An in-stream reservoir acts as a settling basin, and will eliminate or reduce problems of suspended sediment or turbidity. The clearer water may mean lower treatment costs for downstream water users, or it may promote increased algae growth with associated taste and odor difficulties.

Another method for modifying water quality in impoundments (and in natural lakes) is by mixing. The purpose is to destratify the impoundment or to prevent stratification from occurring. A number of installations have been made in the last ten years. Results show that it is possible to destratify an impoundment by mixing and to improve the quality of the entire water mass, but they also show that stratification can reoccur after mixing ceases. Hence, prevention of stratification appears to be the better approach. The desired circulation patterns can be achieved through reservoir designits shape and location of inlets and outlets as well as by mechanical mixing.

The degree to which the assimilative capacity downstream from an impoundment is improved is dependent not only on the reactions which take place in the Impoundment but also on how the reservoir is operated. Depending on the time pattern of releases, there can be favourable or unfavourable effects on the assimilative capacity. For example, when a power plant at a reservoir is used for "peaking," the release pattern will vary substantially. Flows will be high during the portion of each weekday when the peaking load is being met, and low during the night and on weekends. This pattern is in contrast to the more continuous pattern of waste discharges. Reregulation of the power releases can compensate for the varying flow, but of course at increased cost.

Another way of increasing the assimilative capacity of streams and lakes is by artificial reaeration with either air or oxygen. Mobile or fixed aerating devices can be installed in reaches of streams where needed to replenish the dissolved oxygen supply and prevent anaerobic conditions. Reaeration differs from methods for modifying wastes prior to discharge in that nothing is removed from the receiving water, and from methods for increasing the degree of dilution in that there is no increase in streamflow velocity to alter the length and shape of the oxygen sag. The effect is analogous to that of a lengthy oxidation pond artificially aerated. In various coastal areas of the United States the heavy use of groundwater resources has led to saltwater intrusion. One solution to this problem is to reduce the pumpage from the affected groundwater aquifers. Another is to establish barriers to the intrusion. In Southern California, the Manhattan Beach Project for reducing seawater intrusion and preventing further deterioration of the available groundwater supply created a saltwater barrier by a series of injection wells. The well system provides a groundwater gradient toward the ocean which prevents further intrusion and enables the continued use of the groundwater aquifer. Impermeable physical barriers are another way of preventing saltwater intrusion.

Surface water is also subject to saltwater intrusion. Decreased outflow from upstream areas of rivers which empty into tidal estuaries has resulted in sporadic seawater intrusion, as in San Francisco Bay and the Delaware estuary. This type of intrusion can be reduced by increasing the flow into the estuary, or by constructing physical barriers to upstream saltwater movement. The first method has been utilised for many years in connection with San Francisco Bay, where the Central Valley Project has been operated to provide releases to reduce seawater intrusion. The second method has been proposed and studied for both San Francisco Bay and the Delaware estuary. The difficulty with physical barriers is that they impede navigation, affect aquatic life, and aggravate problems of managing wastes discharged upstream. Seawater encroachment in tidal estuaries will become more serious, as upstream development increases the consumptive use of water. Effluent redistribution to make more efficient use of the available assimilative capacity can be achieved in a number of ways including regulated discharge, transfer of the effluent to a different location for discharge, and underground disposal. In general, these methods have little, if any, effect on waste loads as such, and they do not modify the receiving water.

Regulated discharge involves the temporary storage of waste effluents for discharge at times when more water is available for dilution. In an estuary, this may mean storage during the ebb tide and discharge on the flood tide. Often it involves storage of wastes during low-flow periods in the summer and discharge during winter and or spring when streamflow is higher. Some degradation of wastes takes place during seasonal storage, and this procedure is used by a number of foodprocessing, pulp and paper plants, and petroleum refineries. The efficient use of regulated discharge depends on sophisticated studies of the operation of such a system, and on a data collection system that provides continuous information on the quantity and quality of receiving water and the quantity and quality of wastes to be discharged.

Transferring of wastes from one location of potential discharge to one or more other areas where the assimilative capacity is larger is usually practiced in areas near tidal estuaries or oceans. This system is now used by the Toms River Chemical Corporation, which formerly handled the wastes generated in the production of synthetic and organic dyes by equalization, neutralization, clarification, chlorination, aeration, and discharge to the adjacent Toms River. When the residual waste load from growth in production approached the assimilative capacity of the river, ocean discharge was adopted. This made possible the elimination of both chlorination and aeration. The entire waste-handling system is continuously monitored.

In underground disposal, effluents are discharged into relatively deep formations that contain so little water or water of such poor quality that they are not expected to be used for water supply. This method of disposal is used largely by the chemical, petrochemical, and paper industries for wastes that are difficult or very expensive to treat. Underground disposal represents a consumptive use of water, since the effluent is not available for reuse.

The wide variety of measures for water quality management discussed above could be incorporated in plans for a water utilisation system in a region (say a river basin). Planning, however, is only the first step in the management problem. The system must be implemented and once in existence must be operated to produce a specified time pattern of outputs of water and water-related goods and services, including water quality, at specified locations. This it must do in the face of uncertainties relating to hydrologic events and to demands on the water resource over time. Efficient and effective operation of a water utilisation system involves not only day-to-day operation of individual units in the system, such as waste treatment plants and reiteration devices, but integrated operation of all units in the entire system. This is no mean task. Before concluding our comments on the technical and engineering aspects of water quality, we turn to a short discussion of system operation.

Operating a system means determining when to push buttons, turn valves, open and close gates, and so on for the various units involved in, and economic activities related to, it water utilisation system. Such operation requires a communications-control network consisting of data collection, data communication, data analysis and interpretation, signaling, and response, his illustrated in the generalised collection involves continuous recording of streamflow, precipitation, soil moisture, continuous monitoring of various water quality parameters in receiving waters and waste discharges, continuous recording of the volume and quality of water in storage reservoirs and in groundwater basins. In the last decade, considerable advances have been made in the development of instrumentation for continuous analysis of various water quality variables such his dissolved oxygen, temperature, turbidity, pH, total dissolved solids, conductance, and COD (chemical oxygen demand). From the data collection points the information must be transmitted to a control centre. Again, recent developments improved this component of the management system through such techniques is automatic telemetering of data.

In the analysis and control centre, the data are compiled and printed out or stored. Assuming basic relationships have already been developed, such as among streamflow, waste loads, and water quality, and streamflow and stream velocity, estimates can be made of water quality at various points in the system. This information is relevant to both shortrun and long-run situations. Examples of the former are accidental spills of toxic materials and intense storms producing high concentrations of suspended sediment which might clog water intakes or groundwater recharge facilities. Long-run situations involve protracted periods of low flow.

From the control centre this information can be transmitted to various operators and water users in the area. The operators and water users at various locations can then respond to the "signals" received in terms of pushing buttons, turning valves, opening and closing gates. Artificial reaeration devices can be started; releases from reservoirs can be made; additional chemicals added in various treatment processes. Artificial groundwater recharge facilities may be by-passed. If necessary, generation of wastes can be reduced by curtailing production.

Despite all of the problems relating to system operation, the net gains which can he expected from the development of what we have called a communications-control network appear to be significant. Such networks were probably conceived first to preclude serious damage from accidental spills of toxic materials or substances that would disrupt waste treatment and 'or water treatment facilities (e.g., phenols, cyanide, arsenic) or to monitor the performance of waste treatment plants in relation to receiving waters. As demands on the water resource increase, with a concomitant increase iii interactions among various water users, the need for, and the potential savings from, sophisticated system operation will increase. Savings stem from the possibility for more nearly optimal system plans, i.e., reduction in size of some system components, and from the possibility of incremental operation of water quality management systems once they are in existence. The latter refers to the use at any given time of those measures which provide the desired water quality at least cost. The benefits from this type of system operation, including computer control, are well documented in the power industry.

It should be emphasized that the planning of any water management system involves assumptions about the way the system components are to be operated when they are finally In place. Accurate planning *requires* determination of the performance of proposed systems under various possible hydrologic events. Accordingly, to achieve a valid assessment of system performance requires structuring the simulation of the water utilisation system in the planning phase in a manner as close as possible to the situation that will exist in the real world. Even so, the efficiency of operation of system components assumed in the planning phase can be approached, but never achieved, in actual system operation.

The degree of articulation and integration of water quality management system components implied in the above paragraphs has so far not been achieved even in system operation, much less in the planning of actual systems. The Ohio River Valley Water Sanitation Commission, however, has made substantial progress on the operation phase, especially with respect to a communications network. In most regions, water quality management is at a primitive level. At the same time, the technical and analytical means for implementing such systems are developing rapidly. Institutional means lag far behind. Our discussion of the engineering and technological facets of water quality management has simplified many of the processes and relationships, but one central point should be clear. Technologically, there are many alternative measures available to achieve whatever levels of water quality ire desired. Those include various types and degrees of intake water treatment, measures for reducing the generation of wastes, measures for modifying and disposing of wastes after generation, and measures for increasing or making better use of the assimilative capacity of receiving waters.

These measures substitute for one another economically at various rates depending upon the particular waste loads and receiving water bodies involved. Decisions concerning the proper balance among the various measures cannot be made rationally on technical grounds alone, because not all of the benefits from water quality improvement can be measured in dollar terms. Accordingly, values must be introduced into the decision-making process. Further complicating the decision process are both the stochastic nature (uncertainty of occurrence) of hydrologic phenomena and the inadequate understanding (as yet) of the impacts of waste discharges on water quality of various receiving waters and of the effect of various levels and time patterns of water quality on various water uses.

Identifying aternatives for a water quality management system and devising plans which combine them optimally is a demanding task. Moreover, planning and implementation are complicated by the fractionation of decision-making responsibility for the various possible components of such systems. Some measures are within the purview of the individual water users and waste dischargers, some are the responsibility of state, regional, or federal agencies, and some do not fall within the conventional purview of any agency whatsoever. The problem is further compounded by the fact that water quality improvement is generally only one of the desired outputs from a water management system. Both system planning and operation are affected by the complementary and competitive relationships among the various outputs water quality improvement, energy generation, navigation, flood damage reduction, irrigation, water supply, and water-based recreation. We now turn to the economic aspects of regional water quality management.

Withdrawal Treatment Techniques

Methods of contaminant withdrawal include removing or facilitating the removal of contaminated groundwater or soils from the subsurface.

Methods of Withdrawal for Treatment

The feasibility of collection and withdrawal techniques is usually governed by the size and geohydrology of the site. Important characteristics of the geohydrology to be considered include the site surface and groundwater topography, soil hydraulic characteristics, depth to watertable, and depth to impervious layer. Various collection and withdrawal options include the following:

Pumping involves the removal of contaminated groundwater by pumping from wells or drains. Pumping is an effective method for withdrawal which is not limited to plume depth and can also control the lateral and vertical movement of the contaminated plume. The major concern with this method, however, is the amount of time that fluid has to be pumped in order for concentrations of contaminants to reach acceptable levels.

Gravity drains can be used to intercept the plume and utilise the force of gravity to collect contaminants for their removal from the subsurface. This technique also has the ability to control the lateral and vertical movement of the contaminated plume. Unfortunately drains are not very effective in deep aquifers or areas of hard rock.

Withdrawal enhancement uses various methods to enhance the ability to withdraw either groundwater or contaminants from the subsurface. These methods usually involve increasing the solubility of contaminants in water with the injection of steam, solvents, surfactants, nutrients, bacteria, or heat.

Excavation involves the direct removal of contaminated soil and/or groundwater.

Cut-off trenches may intercept the contaminated water, if the plume is not too deep, and drain the water for further treatment or for disposal in a suitable body of surface water.

Treatment Methods

Physical, chemical, and biological methods can be used to detoxify contaminants found in groundwater.

Treatment takes place after contaminated soil or groundwater has been removed from the subsurface, and may be applied at the source, at on-site treatment units, at off-site wastewater treatment facilities, or at the point of use.

The feasibility and performance standards for a treatment method depend on the susceptibility to treatment of the leachate, of the contaminated soil, or of the groundwater. Treatment methods are listed and summarised below; many of these are still experimental and not in general use.

Physical Treatment

Skimming involves the removal of floating contaminants in multilayer solutions (e.g., oil, grease, and hydrocarbons).

Filtration involves the physical retention and subsequent removal of contaminants present as suspended solids.

Reverse osmosis (RO) converts the leachate to fresh water by using pressure in excess of the osmotic pressure of the leachate to force fresh water through a membrane that is permeable to the 284 water molecules but not to the leachate molecules. This process is effective for the removal of most dissolved organics, inorganic salts, heavy metals, and emulsified oils.

Ultrafiltration (UF) also uses a pressure-driven method of separation through a membrane, but it operates at lower temperatures and is suitable only for leachates with larger molecules.

Air stripping. Contaminated water or soil that has been brought to the surface may be air stripped by air injection to facilitate the volatilisation of contaminants and their removal to the atmosphere.

Steam stripping involves the fractional distillation of volatile organics or gases by heating.

Chemical and Biological Treatment

Precipitation, coagulation, and clarification processes precipitate and coagulate dissolved and colloidal particles but are ineffective in removing soluble organic and inorganic substances. Frequently the chemical processes are used in conjunction with other treatment options.

Ion exchange uses a resin bed to remove selected toxic or undesired ions and replace them with harmless ions. The system is mobile and can be loaded with various types of resins (exchange resins or absorptive resins) to demineralize water that is low in organics and to remove organic or inorganic substances. The resins may then be regenerated using acids, bases, or brine solutions. The suitability of various resins depends largely on their ability to be regenerated.

Activated carbon (AC) can remove a wide range of organic and inorganic material from liquid and aqueous-phased streams by sorption on activated carbon (AC) columns (granular) or in suspension in batch reactors (powdered).

Electrodialysis separates and removes positive or negative ions under the action of an electrical field.

Chemical transformation involves oxidation-reduction reactions for the chemical conversion of contaminants to less toxic substances (e.g., by ozone treatment, hydrogen peroxide treatment, ultraviolet photolysis, and chlorination).

Thermal detoxification involves raising of water, soils or contaminants to high temperatures in the presences of oxygen. At temperatures between 1500 and 2000 degrees Fahrenheit essentially all organic compounds are oxidised. Many innovative thermal treatment techniques are currently in the experimental stages of development or recently applied in the field. They include electric reactors and infrared incinerators and 285 use of molten glass and salts to thermally change waste or drive off volatiles.

Biological treatment processes involve the transformation and removal by microorganisms of dissolved and colloidal biodegradable contaminants; they include both aerobic and anaerobic processes by introduced and indigenous microorganisms. Many experiments with anaerobic and aerobic biological degradation, are being researched for both in-situ remediation and groundwater treatment. A treatment process may also include more than one of the above options. For example, biological treatment processing may include an activated carbon step in order to cleanse the contaminated leachate of all types of toxic substances.

Final Disposal

The feasibility of any treatment depends on the amount of endproduct and the method of final disposal.

The three primary final disposal options are discharge to a municipal sewage-treatment plant, discharge to a surface-water body, or land application. Any discharge to a sewage-treatment plant requires pretreatment to remove contaminants that could damage collections sewers, pumping stations, or the plant itself. In order to discharge to a body of surface water, certain planning criteria must be met, such as water-quality standards, federal or state discharge-permit requirements, and overlapping federal and state water-quality laws and regulations.

Land-application processes include slow-rate irrigation, overland flow, and rapid infiltration. The feasibility of these techniques depends on data collected about the decomposition/application rates, toxic effects on crop or volunteer vegetation systems, and soil-renovation capabilities. Often with organics, the method of final disposal is to burn or rebury.

Determination of Suitable Remedial Management Alternatives

Sills et al. (1980) have suggested a two-phase evaluation procedure for determining the suitable remedial management alternatives. The first phase, a preliminary screening, determines the most viable management alternatives for a given site. The use of computer simulation to evaluate potential remedial action 286 options may be considered. In the second phase, these viable alternatives are evaluated according to technical, economic, and environmental criteria. An iterative screening procedure can compare alternatives as to their effectiveness in treating the contaminants, the establishment of their performance criteria, their cost-effectiveness, and their potential environmental impacts.

In determining the cost-effectiveness of each alternative, unit costs, obtained from published guides, can be applied to each activity in the management alternative. These costs will include short-term costs of capital, labor, and installation and long-term costs of maintenance and monitoring. Total costs management alternatives can be compared by converting all capital, operational, and maintenance costs at the end of the life period to an equitable present-worth cost at the first day which, if invested at a given interest rate, would exactly yield the necessary monies to cover annual costs of the management practice chosen. Depending on the selected remedial management alternative and the characteristics of the site, the cost of remedial action may vary from several thousand to several billion dollars. The appropriate response to a groundwater-pollution problem will depend upon the physical characteristics of the site and the nature of the contamination.

At times, no remedial action is necessary or, at least, cost-effective, and some indirect action, such as locating a new water source, is the best solution. In other cases, some remedial action is needed to halt the contamination and, possibly, to rehabilitate the aquifer.

Two categories of management alternatives for remedial action exist: in-situ detoxification, stabilisation, and immobilisation and conventional withdrawal treatment and final disposal. Selection of viable management alternatives will depend on the nature and extent of contamination, site-specific feasibility, including cost-effectiveness, and potential environmental impacts. In any groundwater pollution problem, prevention would have been the best solution.

Table: *Average Costs and Characteristics of Direct Remedial Methods, 1980*

Method	***Average Estimated Costs***[a]	***Characteristics***
Surface Water Control		
Contour grading	510	Increases runoff, reduces infiltration.
Surface water diversion	55	Diverts surface water fromfill.
Surface sealing (Clay, flyash, concrete, PVC)	639-1,336	If locally available, nativeclay is an economicalmeans of retarding infiltration.
Groundwater Flow Control		
Bentonite slurry trench (wall 1,700 ft. long x60 ft. deep)	1,860	Simple constructionmethods; retardsgroundwater flow. Mainlyused for shallow, unconsolidated aquifers.
Grout curtain (wall 1,700 ft. long x60 ft. deep)	3,880	Very effective in permeablesoils.
Sheet piling(wall 1,700 ft. long x60 ft. deep)	2,218	Widely used for shoring.Mainly used for shallowunconsolidated aquifers.

Contd...

Method	***Average Estimated Costs***[a]	***Characteristics***
Bottom sealing (4 ft. deep)	11,000	Leachate collection may be11,000 Leachate collection may beneeded: difficult toaccomplish results underunconsolidated aquifers.
Plume Management [b]		
Drains	64	Effective in lowering watertable a few metres inunconsolidated materials;can be used to collectshallow leachate.
Well point dewatering	514	Suction lift limits depth to20-30 ft.; inexpensiveinstallation; can be usedto collect shallowleachate.
Deep well dewatering	508	Used in lowering deep watertables; high maintenance costs.
Injection/extraction barrier	552	Creates a hydraulic barrierto stop leachate movment; operation andmaintenance costs arehigh.
Chemical Immobilisation		
Chemical fixation of cover	403	Uses chemically fixedsludge to provide a topseal; provides means ofdisposal for sludge; helpsstabilise landfill.
Chemical injection	239	Immobilises a singlepollutant; in most casesnot feasible; result sunpredictable.
Excavation and Reburial		
Excavation and reburial	12,686	Very expensive; difficult construction.

a. For a 22-acre landfill (prices in thousands of 1980 dollars). Original costs updated using theConsumer Price Index.

b. Costs include present worth of 20 years, operation, maintenance, and, where applicable, powerfor a 22-acre landfill.

Source: After M. M. Sharefkin, Shechter, and A. Kneese. “Impacts, Costs, and Techniquesfor Mitigation of Contaminated Ground Water: A Review”. *Water Resources Research*vol. 20, no. 12 (1984), pp. 1771-1783. Copyright (c) by the American Geophysical Union.

Prevention through management, source control strategies, land zoning, effluent charges and credits, aquifer standards, and through guidelines for construction and operation of groundwater-threatening activities would prove to be the most cost-effective alternatives to the very expensive remedial action options.

Aquifer Classification

- Aquifer classification is used by several states as a groundwater management and protection strategy, and is being considered for implementation by other states as well.
- Selective protection of different aquifers to different quality levels would require a classification system. Preservation and restoration of all aquifers to drinking-water use or the prohibition of degradation of present water quality would not require such a system.
- The benefits and the disadvantages of classification systems are discussed.
- State aquifer classification systems in existence are outlined in this chapter and compared. They fall into distinct groups, ranging from simple narrative classification to sophisticated differentiated systems incorporating land use criteria and numerical water standards.

As groundwater becomes an increasingly valuable resource in the United States, strategies for its protection and management are becoming the focus of detailed study in several states. Aquifer classification is one management tool being used by some states and considered by several others. An analysis of the implications of adopting an aquifer-classification system is summarised below, followed by summaries of the Environmental Protection Agency's classification system and actual state aquifer classification systems. State strategies, classification systems, laws, and regulations are further discussed in given paragraph under state and local strategies for managing groundwater problems. The purpose of aquifer classification is to establish water-quality goals for each aquifer and to identify standards or controls necessary to ensure that water quality meets those goals. It is one approach for implementing the state's water policy by formally designating use and water-quality goals for groundwater resources.

A state's choice of policy largely determines the appropriateness of classification as part of its overall groundwater-management program. There are three major policy options: non-degradation policy, limited degradation policy, and differential protection policy. A non-degradation policy aims to protect all groundwaters at their existing quality, and sometimes even calls for their improvement. A limited-degradation policy aims to protect groundwater at as high a quality as possible and to prevent degradation beyond a given quality or standard.

A differential protection policy aims to protect groundwaters of different levels of quality based on current needs, characteristics, and anticipated uses. The first two policies are blanket policies, whereas the last would require the classification of aquifers.

Very rarely has a state adopted a policy that fits perfectly into one of these categories; however, this breakdown provides a frame of reference.As a groundwater protection management strategy, aquifer classification has both benefits and drawbacks. The benefits provided by aquifer classification include:

- Legal protection for valuable aquifers and a basis for siting potential contamination sources in low-risk areas.
- A reduction in unnecessary economic burdens on wastefacility operators by requiring that stringent water-quality standards be met throughout the state.
- A common basis for critical regulatory decisions affecting future use of groundwater resources.
- An opportunity for public involvement in critical policy decisions.
- Guidance for planning and for programs to protect water quality at all levels of government.

The classification of aquifers has the following drawbacks:

- The difficulty and cost of putting the system into place and making it effective.
- The difficulty of delineating the boundaries of aquifers.
- The possibility of serious legal and public-policy limitations on the feasibility of classification. Without the acknowledgment of some form of degradation zones, a classification system could offer very little flexibility in levels of protection given to groundwater areas.
- The difficulty in gaining public acceptance for formally designated waste-receiving zones.

The Process of Classification

No state has an aquifer-classification system in place today that could serve as a model for other states developing classification systems. The variability of geological and hydrological conditions across the country is too great, and the different programs and policies governing the use and protection of groundwater also vary from state to state. Therefore, in order for a state to adopt an effective classification system, many factors, both legal and environmental, have to be taken into consideration.

Definition of Aquifer Boundaries

Groundwater classes can be developed only after some sort of aquifer boundaries are defined. In some areas of the country, distinct aquifer units can be identified that are hydraulically isolated from any other aquifer systems, as, for example, the Cohansey Aquifer in the New Jersey Central Pine Barrens, the glacial aquifer of Cape Cod, the valley-fill deposits located in glaciated regions of New England and intermountain systems of the western United States, and the St. Peter Sandstone Aquifer of Texas. In most areas of the country, however, identification of isolated hydraulic systems is virtually impossible.

There are several ways to define boundaries. The identification of flow systems such as recharge areas, discharge areas, regional flow, and multi-aquifer systems is another aid in defining the boundaries of aquifers. Water quality may serve to delineate aquifer zones. Natural quality is controlled by precipitation, evapouration, geology, and the amount of time the water remains in the groundwater system. As a general rule, the total content of dissolved solids increases with the depth and length of time that the water has traveled through an aquifer from the point of discharge. High evapouration and low precipitation may result in poor quality groundwater. In areas of high precipitation, shallow aquifers may have water of very good quality, with a total dissolved solids (TDS) content of less than 500 mg/l.From a review of the hydrogeologic and water quality data as described above, a state should be able to decide whether or not it can delineate distinct aquifers or parts of aquifers for classification. There may be no method for distinguishing aquifer regions, and in that event they can only be described rather than mapped. Narrative criteria should, however, be used with caution. They are 319 most effectively applied if used to identify regional aquifer characteristics.

Groundwater Classification

Having determined the boundaries of its aquifers, a state can begin to devise a structure for classifying groundwater. Existing systems recognise from two to eight classes. Wyoming has the greatest number of classes agricultural, livestock, industrial (two classes), mining, domestic uses, aquaculture, and "non-usable." The fundamental question to ask is whether or not an aquifer should be designated as being of drinking-water quality. Further breakdowns can be made to reflect land use, existing physical factors, sensitive environmental systems, and other factors. Connecticut is an example of a state where the drinking-water category is divided into three sub-categories based on existing uses and conditions. Classification categories are structured according to combinations of the following factors:

- existing use
- water quality, primarily based on its total content of dissolved solids
- the determination to use either criteria or standards of quality (standards are more specific than criteria)
- land use
- other aquifer characteristics (e.g., soils and geology)
- yield and availability of water regardless of quality (less than 350 gallons per minute [gpm] is considered uneconomical for community water-supply development)
- ability of an aquifer to attenuate and assimilate wastes
- multi-aquifer flow systems (a plan for the management of groundwater resources based on flow systems has been developed for Nassau and Suffolk Counties on Long Island)
- an aquifer associated with mineral deposits and geothermal sources may be classified for production purposes
- contribution to surface waters
- socioeconomic factors
- depth to aquifer.

Classified aquifers may be controlled by numerical standards for water quality, effluent-discharge limitations (based on pollutant toxicity or discharge volume), or non-numerical rules for land-use 320 and waste disposal. EPA's report, "Drastic: A Standardized System for Evaluating Ground Water Pollution Potential Using Hydrologic

Settings," provides a methodology for planners, managers, and administrators that systematically evaluates the groundwater pollution potential of any hydrologic setting from a variety of sources. This system is used by many states in designating classes for their groundwaters.

EPA Aquifer Classification System

In 1984 EPA established an aquifer classification system as part of the comprehensive groundwater-protection strategy. A central feature of the strategy is the establishment of a framework that accords differing levels of protection to groundwater based on its use, value to society, and vulnerability to contamination. EPA has developed guidelines for classifying groundwater that define the classes, concepts, and key terms related to the system and describe the procedures and informational needs for classifying.

EPA's groundwater classification proposes an extensive threeclass system for the management of groundwater (U.S. EPA, 1986a).

The first step in classification is defining the area to be evaluated, the Classification Review Area (CRA). The guidelines specify a CRA as the area within a two-mile radius of a source of groundwater contamination. This segment of groundwater near the source, EPA believes, is most likely to be affected should a release occur (U.S. EPA, 1986a). Information regarding public and private wells, demographics, hydrogeology, and surface waters and wetlands must be collected. A classification decision then can be made, based on the criteria for each class listed below (U.S. EPA, 1986a).

Class I aquifers, or special groundwater, must be highly vulnerable to contamination because of sensitive hydrologic conditions; they must be an irreplaceable source of drinking water, where no practical alternate source is available; and they must be ecologically vital (i.e., they contribute to sensitive ecological systems that if polluted would destroy a unique habitat). Class II aquifers are all other groundwaters that are currently used or potentially available for drinking water and for other beneficial uses. Class III aquifers are those groundwaters that are not considered a potential source of drinking water and have limited beneficial use. They must be heavily saline, with a total dissolved solid content of greater than 10,000 mg/l; or they must have contamination levels above those that can be cleaned up using methods employed by public-water treatment. They must not migrate into Class I and II groundwater, or discharge into surface water and cause degradation. The final draft of the Guidelines for Ground-Water Classification under the EPA Ground-Water Protection Strategy further defines the classes, concepts, and key terms related to classification systems outlined in the strategy. In addition, it defines steps to be taken and the information needed to classify aquifers according to EPA's system.

Comprehensive Strategy to Reduce Waste

Although the requirement of beneficial use without waste is a fundamental provision of western water law, there has been no meaningful enforcement of this requirement in any of the Northwest states. In all four states, the failure to enforce appears to be because the states lack 1) information on actual water use, 2) a clear waste definition, and 3) political support or wherewithal for anti-waste enforcement. The consequences of this failure to enforce include injury to other legal water right holders, both instream and out of stream, as well as harm to the public's rights as the owners of the water resource. In Oregon, at least, there is a recognition of the need to begin enforcing what has been, to date, a meaningless requirement.

In large part, the failure comes from uncertainty about what constitutes waste and fear of the inevitable controversy that will surround enforcement. However, the damage caused by the wasteful practices to rivers, fish, and wildlife and to other out-of-stream water right holders can no longer be tolerated. The states must begin to address waste as part of an overall strategy of restoring streamflows and providing additional water for future human needs. Any such effort should be sensitive to the effect changes in the status quo will have on communities that have come to rely on a certain manner of use. However, these concerns must not be used as an excuse to maintain customary practices and the status quo.

While Oregon is the only major Columbia Basin state to propose a strategy to address waste, neither this strategy nor any of the other states' actions are likely to result in the development of clear standards that will change water management practices in the near future. Moreover, no state has articulated how it will restore streamflows through waste enforcement. Given the water supply crises this region faces, more immediate action is needed. The following recommendations are intended to supply a meaningful strategy that combines enforcement and incentives to reduce waste, increase efficiency, and put meaning into the term "beneficial use without waste." These recommendations urge the states to 1) require all right holders to measure and report water use, 2) develop a clear statewide waste standard, 3) adopt local efficiency standards by category of use, 4) develop and implement a waste curtailment program that includes both mandatory and voluntary measures, and 5) retain control over water recovered from these waste curtailment efforts.

Require Mandatory Measurement and Reporting

The basin states must immediately begin gathering and compiling information on water use and water-use efficiency, with a goal of using that information to create enforceable water-use efficiency standards. It is obvious that the states cannot enforce against waste if they do not know how much water is being used. The simplest and perhaps most important piece of information that states should be gathering is how much water is actually being diverted or pumped by water users. Oregon and Washington have acknowledged that basic information, such as quantifying actual water use, is critical to efforts to eliminate waste and manage the water resource.

All of the Northwest states have the legal authority to require water users to measure and report water use. They should exercise that authority. At the very least, all four Northwest states should require all new users to measure and report. In addition, the states need to develop a strategy to require existing users to measure and report. Interestingly, there is little opposition by the agricultural community (the largest diverter in the Northwest states) to the notion that measurement and reporting of water use is an important tool. Instead, conflict arises in the specifics of any measurement and reporting program.

Water users articulate several reasons for this opposition, including concerns about the state's ability to manage the data that would be produced by such a program and the expense of installing and maintaining measurement and reporting devices. Individual water users are likely, however, to be amenable to a measurement and reporting program in areas of increasing water shortages and disputes over water. In sum, water use information will help to quantify the scope of the waste problem. Such information is critical to enforcement against waste. If a state lacks the resources to undertake a statewide measurement and reporting program, it should at least be implementing a program in target areas. All states should focus on an overall goal of requiring, over time, measurement and reporting of all water use statewide.

Develop Clear Statewide Waste Standard

Each of the major Columbia Basin states should adopt a clear statewide standard for waste. State water management agencies are best equipped to develop and enforce comprehensive waste standards. Unfortunately, these agencies often lack the political will to adopt standards or the money to undertake such a project. If a state lacks the will or resources to undertake rulemaking for basin or subbasin standards, it should at least undertake a program that targets areas of enforcement on a case-by-case basis. While not the preferable alternative, a case-by-case approach will help to set policy and standards for other situations.

While it is obvious that agricultural conditions, crops, and irrigation practices vary widely in most western states, fairness and predictability dictate the adoption of a statewide waste standard. This standard should be "definite [enough] to afford predictability and notice, yet sufficiently flexible to accommodate variable conditions." In addition, the states should be clear that existing uses may be found to be wasteful and that users may be required to change systems or practices as times change. This approach should provide guidance for developing basin or subbasin efficiency standards for specific uses. An analogy can be found in a number of federal laws. For example, broad technological standards have been used to manage water quality under the federal Clean Water Act (CWA). Use of similar standards for waste would help to provide a framework for efficiency standards. The most stringent of the water quality standards under the CWA is "best available technology" (BAT).

The BAT standard requires use of the best technologies that have been or are capable of being achieved by the compliance deadline and that will result in the highest level of efficiencies possible. Cost to the industry is a consideration in development of this standard, but it need not be compared with the benefits of water use reductions.

While a BAT standard would clearly go beyond any existing interpretation of "beneficial use without waste," states could likely justify such a standard in circumstances where the "wasted" water was urgently needed for other instream and out-of-stream uses. This is because waste has been generally defined in terms of whether water use is reasonable and economical in view of other present and future demands on the resource. A second standard, known as "best practicable technology" (BPT), would allow the state to consider a variety of

standard if the user can show that she is replacing an existing process with an innovative technology that has the potential for industry-wide application, that the use of the technology moves towards the goal of eliminating waste, and that it meets one of three requirements.

The second variance, the "fundamentally different factor" variance, would allow users to show that their use is fundamentally different with respect to the factors considered by the agency when it established the limitation. Cost to the user of implementing the technology is not a factor that qualifies for this variance, and the alternate requirement cannot be less stringent than is justified by the fundamental difference.

The third variance, the "economic" variance, allows a user to obtain a modification of the timetable for compliance with the standard if the users show that their economic capability necessitates less stringent limitations, that they are maximising the use of technology within their economic capacity, and that the variance will result in "reasonable progress" toward the goal of meeting the efficiency standard. State adoption of this "economic" variance should require that arguments advanced by the user to justify inefficient use of water (that is, the incremental economic benefits gained by using more water) be balanced against the economic and social benefits of the same water being allocated to other beneficial uses.

The Northwest states should develop these site-specific standards for every type of water user in an area. A state should make it clear that these standards apply not only to all existing permitted, certificated, and adjudicated uses, but also to applications for new uses of water, requests for either certification or licensing of water use permits, applications for transfer of water, and adjudications of water use claims.

Develop and Implement State Waste Curtailment Program

An Aggressive Enforcement Program

The major Columbia Basin states must commit to enforce the "beneficial use without waste" requirement. The states must be clear that their enforcement of the waste doctrine will not be limited to the complaint-based system historically used. Complaint-based enforcement usually only addresses the needs of a few out-of-stream users. It generally ignores the impact wasteful water use has on instream values and on other water right holders in a basin. States should explain to the public that waste enforcement programs are essential for protecting existing water users, the public's ownership in the water, and the state's interest in fostering future economic development.

State Enforcement

Obviously any enforcement program will be limited by budgetary restraints faced by most state water agencies. An enforcement program that targets wasteful use on a watershed or subbasin basis may be a way to get enforcement where it is most needed. If the state is aggressive in its program and targets "hot spots" for enforcement, the message will be sent to water users throughout the state that it is only a matter of time before they will be subject to enforcement. Critical to the success of this strategy is a message to water right holders that, if the state has to expend its resources through an enforcement action, the user will lose any right to use that water the state has determined is waste. This mandatory "forfeiture" of wasted water will give water users an incentive to use voluntary methods to eliminate waste and increase water use efficiency, if these methods can provide some benefits to the user through efficiency improvements (as in Oregon laws).

Citizen Enforcement

The basin states should establish processes to allow citizens to enforce the prohibition against waste. The historic lack of enforcement by the Northwest states results from a combination of political pressure and a lack of adequate information and funding. Thus, there needs to be an alternative to state enforcement.

The states can do this by adopting rules that establish a system for processing citizen complaints. None of the four western states have specific provisions that allow for either administrative or court-based citizen waste enforcement.

This, in combination with the states' failure to develop clear waste standards, makes citizen enforcement actions difficult. While citizen complaints lead to selective enforcement against waste, empowering citizens by providing a mechanism for them to participate in the enforcement process educates citizens as to the waste problem and gives citizens a more direct interest in the issue.

California regulations establish a procedure whereby interested persons can petition the state to request that it investigate alleged misuses of water. If a similar approach were adopted by Idaho,

Montana, Oregon, and Washington, information provided by citizens would help the state determine whether a violation was occurring and would also create more pressure on water users and the state to address the waste problem. In addition, citizen petitions are more likely to focus on damage to the resource than the historic water user complaint-driven system.

Citizen involvement in the process should help build support for the elimination of wasteful use in a manner that protects and restores stream-flows. Increasing citizen awareness of a problem and empowering citizens by providing a process whereby they can request enforcement helps to create the constituency needed for a successful enforcement program. Without a clear and vocal constituency, the state's attempts at waste enforcement, whether through a state-initiated program or in response to citizen complaints, may be stymied by user groups that have maintained political control of water agencies and the western states' legislatures over the years.

Voluntary Alternatives to Enforcement

Voluntary State Approved Conservation Projects

In conjunction with enforcement programs, the states need to make available voluntary programs to increase water use efficiency. A major incentive for water conservation is that the water right holder can retain control of some of the conserved water and use it on lands outside the bounds of the existing water right. Three of the four Northwest states provide users with incentives to increase their water use efficiency. However, only Oregon and Washington allow some of the saved water to be allocated instream. Any conservation program intended to increase streamflows must require allocation of a percentage of the recovered water for instream use. Such a program must also recognise and address the state's interest in protecting other water users. Oregon's conserved water statute contains all of these elements.

In the past, water users have claimed that a conservation program that does not allow them access to more of the recovered water than is currently allowed under the law does not create an incentive to conserve. However, water users will have an incentive to use the voluntary conservation program if the states are actively enforcing against waste. This should include strict timelines for completion of any conservation project, so that entry into the program cannot be used as an indefinite shield from waste enforcement.

State Funding for Conservation Projects

The states should create a fund for water use efficiency projects that will result in water protected instream. The most inefficient users are usually the least able to afford to upgrade their systems. Critical to any program to increase water use efficiencies and eliminate waste is the development of a fund that could be used to help finance water efficiency projects and fund waste enforcement efforts. There are many ways to build. Perhaps the most obvious is to use a combination of state general fund dollars and a surcharge on those that benefit from using the water resource either through a water use fee or a tax on water transfers. Any state monies, or for that matter federal monies, used on efficiency projects must be conditioned to ensure that all or a portion of water recovered be protected instream or, if not needed instream, be available for further appropriation. Washington has a program known as "Referendum 38," which allows WDOE to fund water conservation projects for irrigation districts. WDOE's rules implementing the law allow the state to double the amount of money granted to a district if a complete conservation plan is submitted with the proposal. However, nothing in the statute or WDOE rules requires that a percentage of the water recovered from publicly funded projects be allocated to instream flows, or even to the state.

State Control Over Water Recovered

The states should exert control over the water recovered from waste enforcement actions. Under the laws of all of the Northwest states, in over-appropriated basins, the state has no control of the water recovered from enforcement action.

The water will simply go to the junior water right holders. While this may further state goals of eliminating over-appropriation and protecting existing rights to use water, it will not restore instream flow unless an instream water right has priority. Even though Idaho, Montana, Oregon, and Washington all allow the establishment of minimum stream flows, these flows are usually quite junior in terms of priority. This makes it unlikely that any water recovered from waste enforcement would result in streamflow restoration.

There are two possible solutions to this problem. First, the state could build into its enforcement proceedings a process that allows a water user subject to an enforcement action to enter into an agreement with the state suspending the enforcement effort for a limited period if the user voluntarily pursues water conservation measures under a state law that would allow for protection of some or all of the recovered water instream. The states would have to implement this approach carefully. Otherwise, water users would have no real incentive voluntarily to eliminate waste and increase water use efficiency absent pending enforcement action.

The second approach would be to change state law to require that any water recovered by state waste enforcement action be legally protected instream to the extent necessary to meet instream flow requirements. Such a legislative change should not be subject to a takings claim, because the water resource is publicly owned and the right to use water is and has always been subject to the requirement that the water be used without waste.

A water user who is wasting water (or putting water to a nonbeneficial use) has no constitutional claim arising from regulation by the state. State legislatures can, in response to the increasing demands for water, require state agencies to eliminate wasteful use of water as needed to restore streamflows as a way of protecting the public's ownership interest in the resource.

A potential problem may arise, however, as to the effect of this allocation on junior users. In some instances, a junior user may have come to rely on return flows from inefficient use that was later eliminated by an enforcement action.

Generally, however, a water user has no right to water that is in a stream because a previously inefficient user is now efficient. Another argument can be made that the wasted water should be returned to the river where the next junior in line is entitled to it. However, the state may be able to deny a junior such water if the state determines that beneficial use is maximised by requiring users to meet certain efficiency standards, rather than by seeking to identify and regulate waste.

Although the states are likely to avoid any legal takings problem, it is probably politically expedient for most states to develop a program to mitigate the potential for harm to downstream user. This not only diminishes the chance that the state's action would rise to the level of a taking, it is also equitable. None of the four western states have the tools in place that would allow them to capture the water recovered from a waste enforcement action. If the states' goal is to use waste elimination as a tool to restoring streamflows, the states must take steps to ensure their control of the water recovered. The law in Idaho, Montana, Oregon, and Washington requires that water not be wasted. Although this requirement is a fundamental provision of western water law, there has been no meaningful enforcement of this requirement in any of the states. In all four states, the failure to enforce appears to be because the states lack 1) information on actual water use, 2) a clear waste definition, and 3) political support or wherewithal for anti-waste enforcement. The consequences of this failure include injury to other legal water right holders, both instream and out-of-stream, as well as harm to the public's rights as the owners of the water resource. Enforcement of the fundamental prohibition against waste in western water law could help to solve natural resource and water use conflicts. However, this can only occur if the tools are in place to ensure that the water recovered can be allocated in a way that resolves those conflicts. The states need to adopt a waste curtailment strategy that combines enforcement and incentives to reduce waste, increases efficiency, and puts meaning into the term "beneficial use without waste."

These recommendations urge the states to 1) require all right holders to measure and report water use, 2) develop a clear statewide waste standard, 3) adopt local efficiency standards by category of use, 4) develop and implement a waste curtailment program that includes both mandatory and voluntary measures, and 5) retain control over water recovered from these waste curtailment efforts. State implementation of this strategy will have to be sensitive to the politics of enforcement and to the effect enforcement will have on the other water users as well as to the hydrology of the affected river system.

Just as wasteful practices have changed the hydrology of river systems over time, eliminating those practices will have effects on the system. State sensitivity to these issues is important. However, they cannot be allowed to be an excuse not to tackle waste. When embarking on waste curtailment programs, the states need to be clear on the goal of each program. If one of the goals is to use some or all of the water recovered to protect and restore streamflows, then the states need to make sure that they have the legal tools in place to do this. Otherwise, any public funds, time, and effort spent in this area will themselves be "wasted."

Bibliography

Acharya, Manjushree: *Marine Biology*, International Scientific Pub, Delhi, 2011.

Allan, T.D.: *Satellite Microwave Remote Sensing*, John Wiley and Sons, New York, 1983.

Bagis, Ali Ihsan: *Water as an Element of Cooperation and Development in the Middle East*, Ayna Publications, Ankara, 1994.

Barman, R P : *Marine and Estuarine Fish Fauna of Orissa*, Zoological Survey of India, Delhi, 2007.

Burdak, L.R.: *Recent Advances in Desert Afforestation*, F.R.I., Dehra dun, 1982.

Chauhan, B.S. : *Principles of Biochemistry and Biophysics,* Laxmi Publications, Delhi, 2008.

Curtis, L. F. : *Introduction to Environmental Remote Sensing*, London, New York, 1992.

Diana L.: *People and Pixels: Linking Remote Sensing and Social Science.* National Academies Press, Delhi, 1998.

Elachi, Charles : *Introduction to the Physics and Techniques of Remote Sensing*, New York, Wiley, 1987.

Featherly H. I.: *Taxonomic Terminology of the Higher Plants*, USA, Iowa State College Press, 1954.

Garrison, T.: *Oceanography, An Invitation to Marine Science*, Wadsworth, Belmont, 1993.

Goel, P.K. : *Advances in Industrial Wastewater Treatment*, Technoscience, Delhi, 1999.

Hammer, D.A. and R.K. Bastian: *Wetlands Ecosystems: Natural Water Purifiers*, Lewis publishers, Chelsea, Michigan, 1989.

Husain, Ahmad : *Environment and Water Resource Management*, Sumit Enterprises, Delhi, 2006.

Issar, Arie: *Water Shall Flow from the Rock: Hydrogeology and Climate in the Lands of the Bible*, Springer-Verlag, NYC, 1990.

Jat M.L. and Bhakar S.R.: *Ground Water Hydrology: Theory and Practice*, Agrotech, Delhi, 2009.

Johnson, Cait : *Earth, Water, Fire and Air: Essential Ways of Connecting to Spirit*, Woodstock, VT: Skylight Paths, 2003.

Kamla Devi and D.V. Rao : *Poisonous and Venomous Fishes of Andaman Islands, Bay of Bengal*, Zoological Survey of India, 2003.

Kanmony, J. Cyril : *Drinking Water Management : Problems and P*

Langran, G.: *Time in Geographic Information Systems*. Bristol: Taylor & Francis, 1992.

Lloyd Myers : *Handbook of Water Harvesting*, Washington D.C.: U.S. Dept. of Agriculture, Agricultural Research Service, 1983

Mishra Archana : *Water Harvesting : Ecological and Economic Appraisal*, Authors Press, Delhi, 2006.

Nag, P. : *Geography of Indian Ocean*, National Atlas and Thematic Mapping Organisation, Delhi, 2007.

Pahwa Prem S. : *Water Harvesting, Purification and Distribution Management*, Dominant, Delhi, 2001.

Pandey, Prem Chand: *Advances in Marine and Antarctic Science*, APH, Delhi, 2002.

Rao K. Nageswara : *Water Resources Management : Realities and Challenges*, New Century Publication, Delhi, 2006.

Robinson, I. S. : *Satellite Oceanography: An Introduction for Oceanographers and Remote-sensing Scientists*, New York, Halsted Press, 1985.

Sahoo, Dinabandhu: *Farming the Ocean : Seaweeds Cultivation and Utilization*, Aravali, Delhi, 2000.

Sangha, Prem Chand : *Basic Techniques in Biochemistry and Molecular Biology*, I K International, Delhi, 2009.

Thurman, H.V.: *Essentials of Oceanography*, Merrill, Columbus, 1987.

Todd, David Keith: *Groundwater Hydrology*. John Wiley & Sons, New York, NY. 1959.

Tomi Petr: *Fisheries in Irrigation Systems of Arid Asia*, Daya, Delhi, 2007.

Upadhyaya, Shishir: *Combating Piracy in the Indian Ocean*, Manas, Delhi, 2011.

Vairamuthu. A: *A Drop In Search Of The Ocean*, Rupa Pub, Delhi, 2003.

Wallace, W.J. *Oceanography, An Introduction*, Wadsworth, Belmont, 1995.

Watson, O. Michael : *Symbolic and Expressive use of Space: An Introduction to Proxemic Behavior*, New York, *1972*.